Crop Production
and One Health

Kajal Sengupta
Prasun Chatterjee
Madhurima Bauri

CRC Press
Taylor & Francis Group
Boca Raton London New York

CRC Press is an imprint of the
Taylor & Francis Group, an **informa** business

–EPH–
Elite Publishing House

First published 2025
by CRC Press
4 Park Square, Milton Park, Abingdon, Oxon, OX14 4RN

and by CRC Press
2385 NW Executive Center Drive, Suite 320, Boca Raton FL 33431

CRC Press is an imprint of Informa UK Limited

© 2025 Elite Publishing House

The right of Kajal Sengupta, Prasun Chatterjee and Madhurima Bauri to be identified as authors of this work has been asserted in accordance with sections 77 and 78 of the Copyright, Designs and Patents Act 1988.

Print edition not for sale in South Asia (India, Sri Lanka, Nepal, Bangladesh, Pakistan or Bhutan).

British Library Cataloguing-in-Publication Data
A catalogue record for this book is available from the British Library

ISBN: 9781032822600 (hbk)
ISBN: 9781032822631 (pbk)
ISBN: 9781003503729 (ebk)

DOI: 10.4324/9781003503729

Typeset in Adobe Caslon Pro
by Elite Publishing House, Delhi

-EPH-

Contents

Preface

The subject matter in this book is mainly focused on recognising the relationship between humans, animals, plants, and their common environment in order to attain optimum health results. If we look around, we may witness a variety of daily occurrences that are easily connected with the Newton's third law, "for every action (force) in nature there is an equal and opposite reaction". In this planet each and every living being are interconnected with each other, and stressing one life system can affect them all. This concept has been long recognized both nationally and globally and now it's high time that all disciplines and spheres should come forward and become conscious of their influence on other systems on Earth and working together towards this 'One Health' approach.

The key objectives of this 'One Health' are to identify and track the numerous risk factors, identify the system's strengths and weaknesses that can further reduce the risk factors' rising effects, and to also promote several eco-friendly and more sustainable choices.

Considering all elements of 'One Health' in one location, the four components of 'One Health' have been explained further, which include geographical component describing the geographical spread of the disease; the ecological component, which focuses solely on the environment of our surroundings; the human activities component explaining our positive and negative activities in both the inner and outer framework and lastly the food and agricultural component, covering the food items produced, particularly animal products, which have a significant risk of disease transmission.

Already many international organisations including as FAO, USDA, WHO, and WOAH are consistently exploring, monitoring and reporting the systems in order to identify and avoid any risk bearing factor or emerging and evolving danger. India's effort, on the other hand, cannot be ignored, since both the state and central governments have been collaborating with various sectors such as VCI (Veterinary Council of India), NCDC (National Cooperative Development Corporation), and others even since beginning to tie up loose ends and eradicate negative cycles of these zoonotic diseases.

Numerous obstacles, such as lack of funding and qualified researches, must be resolved in order to tie up loose ends and replace vacant positions. With these concerns, the focus is primarily revolving around several preventative measures that can be put in place to avoid the unexpected eruption of risk-bearing components. Above all, educating people and creating awareness is the crucial key.

Everything in this world comes essentially from the earth, and each of our actions has an impact on our environment and ecosystem, whether directly or indirectly. Agriculture, as one of the most important and dominant sectors and its relation with 'One Health', is a very important aspect. Day by day, agriculture is becoming increasingly tough and this is due to the fact that, in order to feed the world's growing population, we are unintentionally implementing a number of erroneous agricultural methods, such as the indiscriminate use of chemical fertilizers and pesticides, excessive tillage operations and so on, leading us to confront an array of difficulties that influence both our agricultural and health policies, such as climate change, acute hunger and poverty, environmental degradation, the spread of numerous soil-borne illnesses, which further impedes agricultural output, increased mortality rates, and so on.

As a matter of fact, the role of an agronomist is one of the most significant aspects in overcoming these challenges, restoring the neglected areas, and adopting balanced agricultural practises to match the expectations while significantly reducing the negative consequences of agriculture while enriching the contemporary agriculture. A description of how to tackle these unethical activities as an agronomist has been given in this book.

We acknowledge our indebtedness to the authors of many books periodicals, bulletins, etc. from which some of the material that we used here. Information about various 'One Health' approach windows has been supplied by the authors. The book also includes suggestions for connecting the dots and conducting further research.

Generally speaking, it is a very small effort to gather all of the expanding branches of the 'One Health' in the chest of agriculture, recognize and remove the thorns of negligence and boost the twigs of potential alternatives while fortifying the core of Mother Nature. We hope that the book will be of great assistance to students, researchers, teachers, and farmers in understanding the significance of this 'One Health' initiative.

Authors

About the Authors

Dr. Kajal Sengupta (born in 1958), Professor of Agronomy, completed his school level education in 1975 and for his brilliant result in Higher Secondary examination he obtained Merit Prize and Merit Certificate. He got National Science Talent Search (N.S.T.S.) Scholarship (under N.S.T.S. scheme 1975) and enjoyed the scholarship throughout his academic carrier – up to Ph.D. level. He also got Merit Certificate for securing 1st class 1st position in M.Sc. (Ag.) Examination. He is a Fellow of Indian Society of Pulses Research and Development, Kanpur, (at I.I.P.R., U.P.). He got Gold Medal Award of Crop and Weed Science Society (CWSS, Nadia, West Bengal). He is former Dean, Faculty of Agriculture and Head, Department of Agronomy, *Bidhan Chandra Krishi Viswavidyalaya* (BCKV), West Bengal. He had guided a big number of M.Sc. and Ph.D. students. He also acted as the Principal Investigator (PI) of several research projects. He has already published many research papers in various national and international journals and popular extension articles in local papers and magazines / journals. He published more than 25 books and had contributed many book chapters. He regularly participates in various All India Radio (AIR) and Doordarsan (TV) programmes. He has a Professional Experience of more than 38 years. He is a Member of Academic Council / Expert / External Examiner of different Universities / Institutes and also a Peer Reviewer of different renowned journals. After his retirement from BCKV he has joined *Ramakrishna Mission Vivekananda Educational and Research Institute* and presently he is working as Professor and Head, Division of Agronomy, IRDM Faculty Centre, School of Agriculture and Rural Development, Ramakrishna Mission Vivekananda Educational and Research Institute (RKMVERI), Narendrapur Campus, Kolkata – 700103, West Bengal.

Mr. Prasun Chatterjee (born in 1998) completed his school level education from Durgapur, West Bengal. After that, he completed his Bachelor's programme in Agriculture from Uttaranchal University, Dehradun in the year 2021 and is currently pursuing his Master's Programme in Agronomy from *Ramakrishna Mission Vivekananda Educational and Research Institute*, Narendrapur, Kolkata. Currently, he is working on the effect of seaweed based bio-stimulants in coastal ecosystem. He had also authored a book chapter in the book titled "Entrepreneurship in Integrated Farming System". He has a keen interest in salinity ecosystem along with nutrient dynamics of various agronomic crops.

Miss Madhurima Bauri (born in 1997) is a final Semester post-graduate student of *Ramakrishna Mission Vivekananda Educational and Research Institute*, Kolkata - 700103 pursuing M.Sc. (Ag) in Agronomy. She is a meritorious student and loves to read books. She got her B. Sc. (Hons.) Agriculture degree from *Hemvati Nandan Bahuguna Garhwal University*, Uttarakhand in the year 2021. She wrote a few scientific and extension articles, and her latest article "*Joibo projukti tei noya biplob*" got published in *Chasbash* section of **Pratidin** newspaper appreciated by people.

Chapter - 1

Introduction

Crop production is a process in which different types of crops are grown mainly to produce our foods, to meet up our daily demand and to feed our livestock.

The crops, crop sequences, their interaction with farm resources and management practices used on a particular agricultural land over a period of years are very important to sustain the ecology and soil health under a specific environmental condition.

Intensive Crop Cultivation and its Effect

In this type of crop cultivation, which is commonly found in present day agriculture, high quantity of agricultural inputs are used. In intensive agriculture a lot of inputs, money and labour are used to increase the crop yields. Intensive agriculture damages the environment creates / and causes pollution. It is an agricultural system that aims to get maximum yield from the available land.

In present day crop production there is indiscriminate use of toxic agricultural inputs like fertilizers, pesticides and/or other chemical inputs.

The common demerits are:

1. In true sense it is not eco-friendly, increases the pollution level due to the use of a large quantity of commercial fertilizers and pesticides. It has a bad effect on the environment;

2. May kills beneficial insects and plants;

3. Degrades and depletes the soil organisms / soil fertility;

4. May cause genetic erosion of crops and livestock species;

5. May increase the problems of disease pathogens and insect-pests.

There are two crop environments, one is aerial environment (air) and the other is soil environment (soil).

Environmental factors that mainly affect crop production include light (mainly sunlight), temperature, water, air (oxygen, carbon dioxide, etc.), humidity and plant nutrition and it is important to understand how these factors affect crop growth and development.

The main components of aerial environment are sun light (radiation), temperature, rainfall (precipitation) and humidity. Unfortunately we are unable to regulate or control these components and we have to depend on nature and or vagaries of the weather.

Components or factors of soil environment are: soil air, moisture, temperature, nutrients, reaction (pH), biotic factors (mainly microorganisms), etc. Fortunately we can regulate or control these components and crop management depends mainly on how effectively/efficiently or how judiciously we use or manage these components.

Soil environment is the main key to better soil quality and high crop yield.

Soils may also act as a "sink" by storing excess carbon from the atmosphere and, in turn, improving the soil's ability to maintain moisture and nutrients. The same is true of air. When a farmer adds crop nutrients, some of those nutrients are oxidized and lost into the air, increasing greenhouse gas levels. More greenhouse gases mean an overall warmer climate, less crop yield or poor crop quality. Therefore, scientific management of crops is very important to sustain the crop production.

Pesticides used in crop production can affect human health by disrupting hormone and immune system function. Similarly use of insecticides can lead to resistance among vectors of human diseases.

Increasing crop yields through healthy plants is critical for achieving food security for a growing global population. Better and strong plant health can prevent contamination of crops with fungal species that produce dangerous mycotoxins. More use of organic matter (manure), recommended for its beneficial impact on plant health, can reduce the 'One Health' risks.

Health is one. By protecting one we can protect all. Our health is closely connected to the health of soil, animals and our environment (ecology). Now is the high time to make this critical message to deliver to health authorities and society. 'One Health' is not new, however, it has become more important in recent years.

In the year 2003-2004, the term 'one health' was first used and was closely related with SARS. **Recent outbreak of complex health hazards has made it impossible for one sector to provide solution.** The close connection or rather interconnection between human, animal and plant health was glorified in front of our eyes via this pandemic and showed us how animal to human health can be affected by the transmission of the virus. One Health states this interconnection between the health of animals, humans & ecosystems. This overlapping of health of all living beings can be reflected by a venn diagram (Figure 1).

"One Health" may be defined as a trans- disciplinary, collaborative and multisectoral approach which works with the goal to achieve optimal health at the local, regional, national, and global levels through the interconnection between animals, plants, people, and shared environment. One Health work on the areas including food safety, zoonosis control (diseases which spread between humans and animals, such as Flu, Rift Valley Fever and Rabies, etc), and overcoming antibiotic resistance. One health serves with a main purpose of encouraging sharing of knowledge and collaborations in research at multiple levels across disciplines like plants, animal, human; improving soil, environmental & ecosystem health so as to defend and protect the health of all species. Thus participation of various sectors (Veterinary, Medical, Agriculture, Biotechnology, and Extension) and most importantly common people is mandatory for effective implementation of this approach.

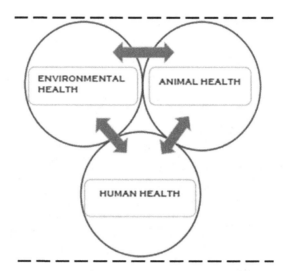

**Figure 1: Interconnection between animal,
human & environmental health**

Human beings can be considered as a main pioneer in deterioration of health of all the system but it had been proven that they are most sufferers and bearers of the greatest harm. For example, some food crops were grown in region where soil health as well as the management practices was not up to the mark thus affecting good yield as well as hampering nutritional quality. This in turn may have a direct impact on the health of consumer and on other hand, if it is fed to the livestock, it will lead to poor animal health and also reduce the quality of products. Another example can be sited based on the usage of Malathion to prevent the maize crops from the aphid attacks which are later consumed by other animals and birds including humans can result in an easy outbreak. Extensive deterioration of ecosystem due to accelerated global changes may be due to the geopolitical problems round the world. There has been omission of the boundaries between the habitats due to migratory movement of others species and mankind, resulting in destruction of stability of ecosystem. Alteration of environment led to the emergence of disease, both infectious as well as non-infectious affect animal, human as well as environmental health. A survey estimated that out of 60% of human pathogens occurrence, more than 75% originated from wildlife. One Health approach not only reduces the potential threats but also protects the biodiversity.

Chapter - 2

History

Many incidents had proven the interdependency of animal, human and environmental health with the existence of various zoonotic diseases such as Avian Influenza (2003), Ebola (2014), ZIKA (2015) and recently emerging COVID pandemic (2019). Before the world grappled with COVID-19 pandemic, the concept of 'One Health' came into existence where Greek Physician Hippocrates (460 B.C. -370 B.C.) in his text 'On Airs, Waters & Places' promoted the concept of public health depended on clean environment. In mid-1800s, Rudolf Virchow came with a term 'Zoonosis' to describe a disease that can be passed from animals to human.

The term 'One Health' was first used in 2003-04, came with emergence of SARS & subsequently avian influenza H5N1 from the previous concept of 'One Medicine' where the veterinary and human medicines were combined to one unit. 'One Health' approach derives its blueprint from Tripartite alliance between FAO of the United Nations, World Health organization & the World Organisation for Animal Health (OIE). Though in earlier term, it addressed zoonotic but gradually expanded to different fields such as ecotoxicology, antimicrobial resistance and urban health, ultimately helpful in potentially achieving the **sustainable development goals (SGDs).**

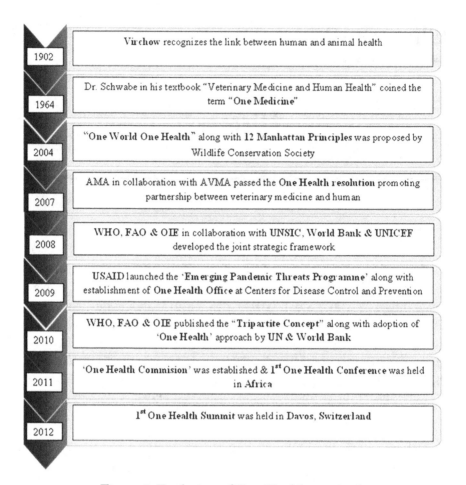

1902	Virchow recognizes the link between human and animal health
1964	Dr. Schwabe in his textbook "Veterinary Medicine and Human Health" coined the term "One Medicine"
2004	"One World One Health" along with 12 Manhattan Principles was proposed by Wildlife Conservation Society
2007	AMA in collaboration with AVMA passed the One Health resolution promoting partnership between veterinary medicine and human
2008	WHO, FAO & OIE in collaboration with UNSIC, World Bank & UNICEF developed the joint strategic framework
2009	USAID launched the 'Emerging Pandemic Threats Programme' along with establishment of One Health Office at Centers for Disease Control and Prevention
2010	WHO, FAO & OIE published the "Tripartite Concept" along with adoption of 'One Health' approach by UN & World Bank
2011	'One Health Commission' was established & 1st One Health Conference was held in Africa
2012	1st One Health Summit was held in Davos, Switzerland

Figure 2: Evolution of One Health approach

Chapter - 3

The Rationale Behind the Concept

» There has been a constant growing up and expansion of the human population into new geographic areas as a result of which people are in close association with domestic as well as wild animal.

» This close association between animals and their environments has led an opportunity for the various diseases to pass from one entity to other.

» Presently, the earth has been constantly experiencing various changes in climate & land use like deforestation, biodiversity loss, agriculture intensification and frequent flooding and droughts and intensive farming practices, which in turn triggers up the process of transfer.

» On the other hand, if we look onto the motion of people, animals as well as animal products, there has been a rapid increment in terms of international travel and trade. So, spreading of diseases has increased across borders and also around the world.

» The whole population of the world is increasing at an alarming rate and is expected to rise from 7 billion in 2011 to approximately 9 billion by the year 2050.

» More than 75% of all emerging human infections have their origin in animals. Zoonotic diseases include Zika Virus, Avian flu, COVID-19, SARS, Yellow fever, Ebola virus, etc.

Chapter - 4

Objectives of 'One Health'

» *E. coli, Salmonella sp., Campylobacter sp.* are most common foodborne infections having origin in animals. In 2005 alone, 1.8 million people died from food borne diarrhoeal diseases.

» Every year 55,000 people die from rabies.

» Thus, according to the global disease experts, One Health is critical approach which can counter the emerging threat against the emerging zoonotic diseases.

» The health professions and their related disciplines and institutions must work together to provide adequate healthcare, food and water for the growing global population.

» The bond between animal, human and plants beneficially impacts the health of all of them.

» The report systems are maintained at the local as well as regional levels to detect & prevent zoonotic disease emergence and thereby control the spread of disease spread.

» Understanding the various risk factors including both cultural as well as socioeconomic contexts for the frequent disease spill over from wildlife entities to humans & domestic animals and also to break the chain compounding the diseases.

» Capacity development at local, regional and national levels for better.

sharing of information as well as co-ordination among institutions.

» Plant health along with veterinary infrastructure **maintenance** by vaccinating farm animals should be dealt in an integrated way with provision of safe food & animal production practices from farm to table.

» Concerning the people about various zoonotic diseases along with their spreading vectors like brucellosis, rabies, plague etc. by integrating them under the 'One Health' frame.

» Now-a-days, AMR or Antimicrobial resistance is an increasing global concern for human & animal health. The capacity of the food agriculture sectors must be increased to combat and risks minimization regarding AMR.

» Food safety promotion at national as well as international levels.

» Promotion of food security at international as well as national level for feeding the increasing world population and fighting against world hunger.

» Practicing good agricultural practices to reduce insect-pest, diseases for ensuring optimum yield for feeding the whole world.

Chapter - 5

Components of 'One Health'

Geographical Component

Since the final 10 years of the previous century, there has been a greater awareness of the globalisation of the commerce in animals and animal-derived products as well as the lowering of obstacles to the worldwide movement of goods. In the recent past, several instances have emphasised the significance of the commerce in animals and animal-based goods in the spread of contagious diseases. Animal commerce has played a role in the introduction of vector-borne illnesses onto free continents in addition to the spread of infections transferred by direct physical contact, such as FMD. In many places, ecosystems are changing due to climate change, especially global warming. As a consequence, the geographic distribution of many vectors is expanding, such as in the cases of WNF, blue tongue, RVF, dengue, and malaria.

Population densities—both human and animal—are important determinants of the heightened transmission, amplification, and spread of newly developing diseases. The quadrupling of poultry production in China over the past 10–15 years has had a substantial effect on the onset and dissemination of H5N1 HPAI. The mobility of humans often and widely, as well as the booming commerce in unsanitary animal products and by-products, all contribute significantly to the development of (re-)emerging serious infections. In terms of public health, the spread of the novel H1N1 influenza virus and SARS through air travel was powerful symbol of the dangers posed by the migrations of infected hosts.

The awareness that nations are now increasingly linked has led to a shift in worldwide disease management policy. More efficient and effective systems on animal and animal tracking based on ongoing data access among shippers would enable health adaptation in the event of a crisis as well as the deployment of resources required to block the initiation of infectious substances in a more appropriate manner

Ecological Component

Man is frequently an unintentional host of many zoonotic illnesses (including West Nile fever, Rift Valley fever, brucellosis, listeriosis, yersiniosis, trichinosis, and hydatidosis), and he does not contribute to disease transmission (he is referred to as a 'dead-end' host). The involvement of resurrected hosts is sometimes tied to cultural behaviours rather than human biology or pathogen biology. The term "reservoir" is frequently misapplied when discussing wild animals. An alive or inanimate item on or in which a microbe generally dwells, and which, therefore, is frequently a source of contamination by the agent is what the term "reservoir" denotes (Thrusfield, 1995). In some other terms, a reservoir is just a particular kind of host that has the capacity to sustain an infection in the wild, allowing the pathogen to survive throughout the inter-epidemic intervals. They now play a unique function in the infection's epidemiology as a result. In this paradigm, wild boars, though they may contribute to the geographical spread of the virus, may not be regarded as carriers of classical swine fever (CSF) in Europe.

The importance of the urban and suburban areas environment in the transmission dynamics of serious infectious diseases and zoonotic disease should be brought to light. WHO (1988) defines an urban area as: 'a man-made environment, encroaching on and replacing a natural setting and having a relatively high concentration of people whose economic activity is predominantly of the non-agricultural kind'. Some zoonoses have major aspects transmitting in urban environments, which may enhance pathogen spread, such as water-borne illnesses in cities with poor sanitary standards or the spread of particularly infectious person-to-person infections in densely populated metropolitan regions.

Human Activities Component

Medicine remained a single field until the mid-eighteenth century, when veterinary institutions were established in Europe to meet the demand for more skilled operators (Mantovani, 2008). Lyon (established in 1762) became the first veterinary medicine faculty, followed by Turin (1769). Over the last 30 years, WHO has promoted strategies for greater collaboration between the medical and veterinary sectors, giving rise to the concept of 'Veterinary Public Health' (VPH), which was originally defined as 'a component of public health activities devoted to the application of professional veterinary skills, knowledge, and resources to the protection and improvement of human health' (WHO, World Health Organization, 1975).

One of the key priorities of Italian veterinary administration after WWII was the rehabilitation of public slaughterhouses in order to repair the meat supply chain. Veterinary and medical services were originally included into the High Commissariat for Hygiene and Public Health in 1945, and subsequently, on March 13, 1958, under the newly formed Ministry of Health. This shared identity permitted the development of a preventative strategy encompassing all these global health and resources. This strategy was also used by the European Commission, which, in the aftermath of the BSE crisis in 1998, transferred animal health competencies from the agricultural sectors to the Animal Health and Consumer Protection Directorate (DG-SANCO). VPH has an interdisciplinary scope, including It includes not only veterinarians from the government, non-government, and commercial sectors, but also other professions also including physicians, nurses, microbiologists, environmental scientists, sanitarians, food scientists, agricultural scientists, para-veterinary workers etc., and auxiliaries who make a significant contribution to the treatment, control, and prevention of illnesses of animal origin (WHO, 2002). This necessitates an active flow of information among experts from many sectors and stakeholders, with the following factors in mind:

1. The development of connections among control and research centres should be expanded to include a wide range of capabilities capable of producing new information.

2. Various scientific funding programmes and international organisations, such as the OIE, promote and support this strategy, as well as the

creation of risk effective communication at the international, national, and local levels.

3. Risk management methods for public health should reflect the values of the various publics to whom they are aimed. The challenge for all those working in the field is to investigate options for involving key stakeholders throughout the process, while ensuring that the kind and level of engagement reflects the breadth and effect of the specific risk situation in question.

4. Also, this calls for strong communication that is based on an understanding of the many varied viewpoints and values that each party engaged has.

Food-Agricultural Component

Consumers in industrialised nations are calling for a complete and integrated approach to food safety (the so-called "farm to fork" approach), which has implications including both producers and regulatory agencies (European Commission, 2000).Control authorities must also be able to carry out appropriate risk assessments to identify and quantify hazards across the food chain and execute streamlined and efficient risk management programmes. The goal of "farm to fork" legislation is to ensure that animal products sold for human consumption are of the best possible quality. 'Quality' means not only the absence of pathogens and it is not a simple synonymous of 'safety', but it is defined as 'the totality of characteristics of an entity that bears its ability to satisfy stated and implied needs' (ISO, 1995).

Any preventative measures must start at the very beginning of the food chain, particularly feed supplied to animals, taking into account the hazards to which individuals are susceptible at the top of the food chain. The occurrences related to the usage of feed contaminated with dioxin (Bernard et al., 2002) or mycotoxins (European Commission, 2009) are an easy way to confirm that the security of plant health is an important aspect of this method. Implementing a successful "farm to fork" policy can often be difficult due to divides between before- and after-harvest competencies. To provide an integrated surveillance network throughout the entire production chain, including that of monitoring its effects on human health, multidisciplinary and multi - sectorial techniques should be undertaken.

Chapter - 6

Efforts made by Various International Organisations

At the global level, four organisations- World Health Organization, Food and Agriculture Organization, World Organisation for Animal Health, and the United Nations Environment Programme have joined together to work out strategies as far as the inter-relatedness and the way to move forward for One Health are concerned.

1. FAO (Food & Agriculture Organization)

FAO's priorities include:

» Monitoring, surveillance and reporting systems at the local, regional and national levels to detect and prevent animal as well as zoonotic disease emergence and spreading of the disease must be strengthened.

» Elimination of hunger, food safety & healthy diets at national and international levels to protect the farmers from the impacts of plant and animal diseases along with increment in the resilience and sustainability agriculture-food ecosystems. A key priority programme area is integration of One Health in agri-food systems transformation & is also a part of FAO's Strategic Framework (2022-2031).

» For combatting and minimizing the risks of AMR, the capacities of the food and agriculture sectors must be increased.

» To prevent and manage disease outbreaks, risk factors for spill over of diseases from wildlife to domestic animals and then to humans including cultural and socioeconomic contexts must be understood.

» For better coordination and information-sharing among stakeholders and institutions, capacities to be developed at local, regional and national levels.

» Veterinary and plant health infrastructure must be reinforced along with safe practices for food and animal production from farm to table.

» At the human-animal-plant-environment (HAPE) interface, there is a need of improvement in the early warning systems on plant pests and diseases & animal, including zoonotic diseases along with build-up of strength for pest and disease management in plants and animals and invasive alien species management.

On 11th January 2021, One Planet Summit was hosted by the President of France, Emmanuel Macron, where the discussion started on combatting the climate change, followed by raising the level of ambition by international communities along with responding to the questions raised by COVID-19 pandemics, thereby recognizing the need and importance of "One Health for all".

2. U.S. Department of Agriculture (USDA)

» **Solving problems associated with antimicrobial resistance:** Antibiotics can be termed as lifesavers. Treatment of various bacterial infections and diseases such as bronchitis, pneumonia, and strep throat, infected wounds and ear infections need antibiotics as the only cure. Producers and veterinarians keeping the food supply safe move towards more judicious use of antibiotic in food animals. But due to rapid use, certain bacterial strains used to treat infections in animals and humans have become resistant. USDA recognizes the ability of bacteria and other microbes to proliferate and survive the effects of an antibiotic or antimicrobial resistance (AMR) is a serious threat to both human health and animal health.

» **Buzzing into action to support pollinator health through research:** Pollinators plays an important role in agricultural production in pollination services for the fruits, nuts and vegetables which contribute to a healthy diet. Pollinator Research Action Plan proposed by USDA tells us about the education, research and economics that supports the national strategy through various extramural and intramural research programs of promoting the health of honeybees and other pollinators. Commercial blueberries are pollinated by an effective pollinator bee, Osmia ribifloris on a barberry flower.

» **Going wild about water at the World Water Forum:** The most important and precious resource is water and increasing human population and their irrational use will lead to its scarcity soon. In many areas, weather patterns are highly affected by climate change. USDA showed the latest science and technology research that is being done at the department at the 7th Annual World Water.

Open Data: - a key to feeding 9 billion people by 2050: According to USDA, open data for agriculture and nutrition could be an essential key for harvesting enough crops to meet future challenges on food scarcity affecting livelihood. **GODAN** (Global Open Data for Agriculture and Nutrition (GODAN) partners at the 3rd Annual meet met to discuss on broadening the partnership in 2015. GODAN is a great initiative which aims of making global efforts support to make nutritional and agricultural data available, usable and accessible for use which in turn unrestricted worldwide.

3. World Health Organization (WHO)

» A One Health Initiative was initiated by World Health Organization with the purpose of integration of animal, human and environmental health across nations.

» World Health Organization also collaborated with the Food and Agriculture Organization of the United Nations (FAO), the World Organisation for Animal Health (WOAH) and the United Nations Environment Programme (UNEP) as a One Health Quadripartite. This collaboration also includes working with political leaders for establishment of the needed infrastructure and channelization of funding. The main role of the quadripartite is the promotion of multi-sectoral approaches for reduction threats related to health at the human-animal-ecosystem interface. The Quadripartite One Health Joint Plan of Action (OH-JPA) basically outlines the transformation that are required to mitigate the effect of the current as well as future challenges to health at global, national and regional levels.

» On May 2021, panel named One Health High-Level Expert Panel (OHHLEP) was formed to advise WHO, FAO, WOAH and UNEP on issue related to One Health. The primary task of the panel is to conduct research on emerging disease threats, along with development of a long-term global plan of action to prevent the outbreaks of diseases like H5N1 avian influenza, MERS, Zika, Ebola and possibly, COVID-19. It will also perform investigation on the impact of human activity on the wildlife habitats, and environment and observe how the disease threats are triggered upon this act.

» Integration of One Health across its different offices and units for providing strategic advice related to policy and conduction of training at the national, regional and local levels.

4. OIE (Office International des Epizooties) or WOAH (World Organisation for Animal Health)

Office International des Epizooties (OIE) was founded in 1924, later converted in WOAH during the year 2003 (May, 2003), is an intergovernmental organisation which focuses on improvement of animal health globally along with maintaining transparency in disseminating information on animal diseases, thus building a healthier, safer and more sustainable world.

» One Health High-Level Expert Panel (OHHLEP) was formed to advise WHO, FAO, WOAH and UNEP on issue related to One Health, which provided a better definition explaining how disciplines, sectors as well as society connect as a whole with four main pillars: collaborations, communications, coordination and capacity building.

» Priority areas include antimicrobial resistance, avian influenza and rabies, established in the year 2011. Others include reduction in the risks emerging from the zoonotic epidemics and pandemics, strengthening of management, assessment & communication of risk related to food safety.

» Collaborations between animal and human sectors have often neglected the integration of the health of the wildlife sector to prioritization as their health directly links to the chain of disease surveillance. With the protection of the wildlife, the organisation safeguards the biodiversity. After the outbreak of diseases like Ebola or COVID-19, the role of wildlife, wildlife trade and ecosystem as a whole can't be overlooked.

» Early warning of an outbreak is the stage where work has been carried out by WOAH. It helps in mitigating the potential threats via communications between various sectors, which gains call for the concept of One Health. GLEWS is a global platform where advice is provided on the prevention and control of health threats or events of potential concern affecting the human-animal-ecosystem interface.

» After the COVID-19 emergence, global leaders united along with WOAH which led to the3 conclusion of building an international pandemic instrument that would protect the planet from upcoming crisis. WOAH carries the voice of health of animals via participating in the negotiation processes led by International Negotiating Body or INB, coordinated by WHO in support of development of new pandemic instrument which functions by not only strengthening of global collaborations but also guarantees political commitment at times of pandemic arousal.

» WOAH in partnership with WHO launched an interactive IHR or PVS National Bridging Workshops (NBWs) with an objective of targeting a fully operational One Health approach at the national level.

The three major international organizations, **FAO** (Food and Agriculture Organization), **WHO** (World Health Organization) and the **OIE** or **WOAH** (World Organization for Animal Health) are working together to control health risks at the ecosystem level. Development of global strategies and tools for ensuring a harmonized and consistent approach throughout the world and better coordination of human, environmental & veterinary health policies at the international and national levels are of much significance.

Other organizations that have taken initiatives regarding one health are:

» American Veterinary Medical Association

» American Medical Association

» Centres for Disease Control and Prevention

» United States Department of Agriculture

» Vétérinaires sans Frontières/ Tierärzte ohne Grenzen

» United States National Environmental Health Association

[The need to fight animal diseases at global level led to the creation of the Office International des Epizooties (OIE) through the international Agreement signed on January 25th 1924. In May 2003 the Office became the World Organisation for Animal Health but kept its historical acronym **OIE**].

Chapter - 7

India's Efforts on Advocating 'One Health' Approach

Now-a-days, One Health approaches are increasingly focussed by both Central and State governments along with various agencies to tackle rapidly emerging issues of zoonoses, food safety and antimicrobial resistance. Few recent development or approaches undertaken concerning the One Health approach are:

» **2007:** A National Standing Committee on Zoonoses was established in the year 2007 keeping in view the impact of zoonotic diseases on public health in India.

» **2009:** India made collaboration with One Health network of South Asia (OHASA), which was formed by the Wildlife Trust to develop a cohesive regional network, consisting of policymakers from Ministries of Health, Agriculture and Environment as well as scientists along with various Non-governmental Organizations and universities in India, Pakistan, Bangladesh & Nepal to counter the rising zoonotic diseases.

» **2010:** Various OH collaborations were formulated between National Centre for Disease Control [NCDC] & Centre for Disease Control and Prevention [CDC], for facilitating and encountering emerging global diseases across the world. A Global Disease Detection – India Centre was established in NCDC, which is a joint collaboration between MoHFW, Government of India and the US Department of Health and Human Service, CDC. A program named Intersectoral

Coordination for Prevention and Control of Zoonotic Diseases is also ran by NCDC which enables us to strengthen the coordination between Veterinary, Medical, Wildlife Sector & and various relevant stakeholders.

» **2014:** A new One Health (OH) centre under the name of Centre for OH Education, Advocacy, Research and Training (COHEART) was established by Kerala Veterinary and Animal Science University, India on February 26, 2014 with inclusion of two new courses in OH, namely, PG Certificate in OH Surveillance & PG Diploma in OH. KVASU also aims to develop OH professionals for fighting against pandemic through their PhD programme in One Health.

» **2015:** For bridging the gap between the two sectors, an employment initiative was taken to recruit cross-sectorial professional starting from a veterinarian in each of the states in India at the IDSP State Surveillance Unit along with a veterinary consultant, who is being recruited by different state government's Directorate of Public Health.

» **2016:** The Veterinary Council of India under the powers conferred by Indian Veterinary Council Act, 1984, for the first time introduced the OH concept in B. V. Sc.as well as in the M.V. Sc. Course.

» **2017:** In the year 2017, MOHFW, Government of India under three governance mechanisms prepared its first National Action Plan on Antimicrobial Resistance (NAP-AMR), taking a One Health approach in the animal, human & environmental sectors.

» **2018:** A path breaking initiative namely, IDSP's Integrated Health Information Platform was launched on November 26, 2018 by Union health secretary, which was built to monitor public health surveillance. This platform will provide data near-real-time for detection of outbreaks, helps in reducing morbidity and mortality, lessens disease burden in the populations with a better health systems to policymakers.

» **2019:** "Delhi Declaration" was signed by Health Minister of India along with other Member States on September 03, 2019, aimed to support, develop and implement intersectoral coordination mechanisms based on the One Health approach. Along with it, India also launched 'One Health Initiative' together with support from Bill & Melinda Gates Foundation, US Defense Threat Reduction Agency & Biological Threat Reduction Program & Penn Skile's Applied Biological & Biosecurity Research Laboratory (ABRL).

» **2020:** Indian Council of Medical Research (ICMR)-National Institute of Virology, Pune in collaboration with Maharashtra Animal and Fishery Sciences University developed One Health Centre in Maharashtra. The central government proposed a national institutional platform while announcing recovery plans for COVID-19. Prime Minister Atmanirbhar Swasth Bharat Yojana{PM-ASBY} supported by ₹3500 cr loan from JICA against COVID -19 in India.

» **2020:** To promote the trans-disciplinary multisectoral cooperation & collaboration for achieving OH framework in India, establishment of a National Expert Group on May 14, 2020 along with Integrated Public Health Laboratories were conducted. Funds were allocated for the setting up of the *Centre for One Health at Nagpur.*

» **2021:** In India, 1st consortium on 'One Health' was launched by Department of Biotechnology. This Consortium majorly consisted of 27 organisations led by DBT-National Institute of Animal Biotechnology, Hyderabad followed by AIIMS Jodhpur, AIIMS, Delhi, GADVASU, Ludhiana, IVRI, Bareilly, MAFSU, Nagpur, Assam, TANUVAS, Chennai, etc. Various veterinary & agricultural universities with many more ICAR, ICMR centres & wild life agencies were also part of the consortium. Recently, Food Safety and Standard Authority of India (FSSAI) made up collaboration with National Centre for Disease Control (NCDC) & National Standing Committee on Zoonoses to work on the various policies and regulatory mechanisms. A manual for handling various zoonotic diseases was published by the Centre of Zoonosis, NCDCRs. 13,343 crores have been sanctioned for Foot and Mouth disease and Brucellosis control under the National Animal Disease Control Programme.

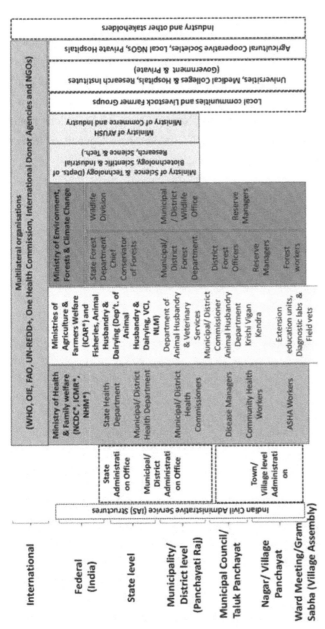

Figure 3: A simplified illustration of the various sectors and the politico-administrative head of the health system in relation to zoonotic disease control in India.

[**Source:** Asaaga *et al.*, 2021]

[ASHA=Accredited Social Health Activist; FAO=Food and Agriculture Organisation; NCDC=National Centre for Disease Control; ICMR=Indian Council for Medical Research; NHM=National Health Mission; ICAR=Indian Council for Agricultural Research; NLM=National Livestock Mission; NGO=Non-government Organisation; OIE=Office International des Epizooties (World Organisation for Animal Health); UN-REDD=United Nations Reducing Emissions from Deforestation and Degradation Programme; VCI=Veterinary Council of India, WHO=World Health Organisation]

Chapter - 8

Scope of 'One Health'

One Health approach urgently need attention to major areas including all levels of academia, industry, government, policy and research, because of the interconnection between that of plant, animal, human, environmental, and health of planet. The areas of attention are:

1. Agricultural land use and food production

2. Effects of climate change and its impacts on health of animals, humans and ecosystems

3. Mitigation of antimicrobial resistance

4. Disaster management

5. Soil plant / health

6. Establishing commonality of diseases among people and animals such as cancer, obesity, and diabetes i.e. comparative medicine

7. Food security & safety

8. Water security and safety

9. Conservation of natural resources

10. Prevention and response of both chronic and infectious (zoonotic) diseases as well as disease surveillance

11. Environmental health

12. Global commerce, trade and security

13. Vector-Borne Diseases

14. Public policy and regulations

15. Communications and outreach

Chapter - 9

Challenges

» There is always risk of more than 1.7 million viruses circulating in wildlife which have the potential threats to human health. We, humans have a very little knowledge about the normal parasite fauna of wildlife (Mathews, 2009; Thompson *et al.*, 2010). Whatever we know, much of it is fragmented and obtained as part of targeted investigations. We know little about key threat the Indian subcontinent possess being a hotspot for vector borne pathogens, zoonotic, drug resistant diseases with no mentioning of 'emerging infectious diseases' and 'zoonoses' by National Health Policy has been noticed. So, it has become increasingly important for us to understand the transmission & impact of parasites in various wildlife forms. The regulation of wildlife populations is often influenced by the negative effect of the parasites in addition to the zoonotic potential (Watson, 2013) and are considered to be the primary cause for the decline of the wildlife species (Abbott, 2006; Wyatt *et al.*, 2008; Robinson *et al.*, 2010; Cameron *et al.*, 2011; Ewen *et al.*, 2012; MacPhee and Greenwood, 2013; Wayne *et al.*, 2013). We need to better understand how environmental changes affect the parasite- host relationship in terms of susceptibility as well as diseases (Harrington *et al.*, 2013).

» For detection of antimicrobial resistance and consumption in animals, human & environment, there is absence of a fully standardized surveillance system (WHO, 2014). As a result of which, no quality data can be generated due to lack of harmonization and standardization. Out of total, only 30% of surveillance program is coordinated and monitored by WHO (WHO, 2012). Moreover, there is still data insufficiency where we can compare the antimicrobial resistance amongst livestock, humans, fishery, poultry, agriculture and environment.

» It has been found that in the European countries, standard guidelines of EUCAST are followed by only 64% of the participating laboratories (ECDC-EFSA-EMA, 2015).

» There is a huge lacking in substantial amount of funds which is required to meet up the training costs of the staffs involved in interdisciplinary work, costs for collection, analysis and sharing of the data results, expert consultant's cost along with increase in the surveillance program coverage across animal, human & environmental sectors (Queenan *et al.*, 2016).

» There is a shortage in veterinary manpower. Inspite of 46 veterinary colleges and 460 medical colleges in India, most of these colleges do little or no research.

» In developing countries like India with high & middle income group, there is lack in the infrastructure facility starting from poor transport facility leading to delay in the quick sending of biological samples into the laboratory, inadequate laboratory setup for the generation of primary data at the local level along with recruitment & training of staffs.

» One of leading challenge in One Health approach is the ethical approval for interdisciplinary one health research work as multiple institutes are involved with variable ethical parameters from different countries (Mfutso-Bengu and Taylor, 2002).

» Other challenges which arise in implementation of One Health programme include lack of active citizen forum, rapid changes in policies of the government, corruption performed by various agencies, lack of proper governance with most importantly lack of respect & trust for human rights, etc. There is a problem in coordination between inter sectorial and government structure along with animal, human and environmental health being controlled by various ministries.

» Poor coordination is observed at slaughter houses for distribution, and retail facilities on area of food safety.

» Antimicrobial resistance (AMR) is considered to be a complex and multifaceted problem which not only threatens humans as well as animal life and health, the global as well as national economy and demands a holistic approach for its mitigation which emerge to be an definite challenge to the researchers across fields.

Chapter - 10

Preventive Measures

There is a famous quote by Benjamin Franklin, **"An ounce of prevention is worth a pound of cure"**. Besides, it is evident that all systems are interconnected with each other. So, to prevent the domino effects, prevention is the only way no matter how small the system is. Some of the preventive measures which need attention are:

1. Since, humans are the spearhead of all the systems, the easiest way to prevent the facing obstacles is educating them. The importance of education in linking environment and health cannot be overstated, especially considering the increasing awareness of the need for worldwide incorporation of One Health for a healthier and safer planet. Hence, education should begin at all levels of society, including higher academic levels, all levels of government, and among professions.

2. Strengthening the national and international emergency response capabilities to prevent & control disease outbreaks is another most vital point of preventive measure that needs to adopt ASAP. Several agencies like WHO, FAO, OIE, UNSIC, World Bank & UNICEF – all are working together, taking their small steps further to make this approach a successful one.

3. Shifting focus from developed to developing economies as well as countries for better addressing the concerns of the needy people is utmost important. The broader the centre of attraction, the more coverage of the approach.

4. More preventive action at animal-human- ecosystem interface which includes maintaining proper distance while interacting with one another, proper sanitation, etc. Adopting appropriate measures helps us to take our one small step towards this global collaboration.

5. Promotion of institutional collaboration across various sectors and disciplines need to be urgently addressed. This is a necessary part as it easily catches public awareness and becomes easy to convey a small message to a large group of entity.

6. Regarding the management of waste or even air or water pollution, collaborative action is the need as much of the marine ecosystem is being threatened due to untreated sewage wastes and industrial effluents discharge, thus ultimately affecting the sustainability of public health & living resources. We can't address the control of vector-borne disease without considering the impact of man-made as well as natural environmental changes in terms of proliferation of disease vectors.

7. Another areas where utmost attention need to be drawn is regarding food and prevention of food safety hazards which include surveillance as well as research of foodborne pathogen, minimizing the food contamination by infectious pathogen, investigation of foodborne illness outbreak, laboratory based research as well as network for identification of the pathogen and conduction of various demonstration and training programs on measures related to food safety.

8. Adoption of a model for an Integrated One Health surveillance system along with various preventative mechanisms amongst the relevant stakeholders for regular and emergency coordination.

9. Development of a standardized approach for identifying the various risks of spill over of pathogens between humans & different animal populations along with the emergence of zoonotic diseases, particularly those that arise in food systems.

Priority areas should include:

a. Integrated disease surveillance;

b. Joint outbreak response;

c. Targeted R&D for gaps in disease preparedness starting from vaccines, diagnostics, therapeutics and platforms;

d. Preparation for pandemics, both human and animal;

e. Streamlining the regulatory aspects in 'One Health';

f. Improved mechanism for data collection, analysis and sharing;

g. Linking appropriate activities to global efforts;

h. Strengthening the routine prevention programmes.

Chapter - 11

Relation of 'One Health' with Agriculture

Plants are the foundation of the energy pyramid, and thus no food web is fully functional without them. Maintaining good health of plants in turn will help to maintain good health of the environment. Once environmental health is maintained, it will be a step forward towards achieving one health goal. The population outburst has posed immense challenge to the agriculturists especially the agronomists to increase production by several folds in order to feed the growing population.

The present world population is at 7.6 billion and in year 2050, it is predicted to be 9.2 billion. To feed the growing population by 2050, global agricultural production must be boosted by 60-70% from current levels. Moreover, increased economic growth and income levels have driven people to consume more meat and dairy products. This change in dietary habits has led to production of more fodder crops for feeding animals under intensive feeding systems. So, efforts should be driven towards production of both food and fodder crops. It is seen that the per capita availability of arable land was 0.42 ha in 1960 which will be reduced to 0.19 ha by 2050. Increased urbanisation, erosion, faulty agricultural practices, deforestation, pollution, -all this has led to gradual decrease of arable land. So, in order to increase production with decreasing land intensive cultivation practices are to be taken. Again, we are aware that even more food is required to feed our expanding population. But agriculture already contributed 34% of the world's greenhouse gas emissions, uses 75% of our freshwater and is the largest driver of biodiversity loss, especially for birds. Therefore, the problem is to raise the level of productivity to feed the world without escalating agriculture's negative environmental effects. Fixing agriculture is the key to human and planetary health.

At present there is a shortage of agricultural lands to meet up the huge demand of food and for this intensive farming has been adopted in many areas which includes the use of higher inputs (particularly use of chemical fertilizers is increased) in order to obtain higher outputs. However, the use of excess and imbalance amount of fertilizers into the soil leading to more residues which can easily destroy the soil microbes and other organisms living in the soil decreasing the soil biological fertility and crop yield with time being and increasing weed growths and populations as well as insect-pest attack. Again, weed management / removal process and controlling of insect-pest attack requires increased use of herbicides and pesticides. Upon washed by irrigation water or rain water, these chemicals as well as residues get leached out with the surface runoff and mixes into the river where it results in eutrophication causing an increase in biological oxygen demand which further leads to nutrient pollution, water pollution and creating the disturbances in the aquatic lives. When livestock animals like cow, buffalo drinks that water heavy metals like cadmium also gets accumulated inside their body which results in milk contamination; ultimately affecting the people who consume dairy products on a regular basis.

In a broad spectrum, this "One Health" approach is also an effective means for preventing the spread of pathogens and limiting antimicrobial resistance in plants and animals, which can negate the effect of important mediums to ensure food safety. One apparent strategy to reduce antimicrobial resistance is to use of less antibiotics in cattle. In actuality, livestock receives 73% of all antibiotics used worldwide. To enhance other agricultural methods and keep cattle clean and healthy, its usage must be reduced. Agriculture that helps to sustain a large population has also increased public health and environmental risks. Agriculture has been linked to bad health issues as a result of livestock-related illnesses, starvation, food-borne sickness, and other factors.

Reduced global consumption of healthy foods has become key dietary contribution of major cause of malnutrition and poor health leading to low economic productivity and death. Consumers now face relatively high prices for agricultural goods due to excessive waste and substantial post-harvest losses that have constrained domestic production. Several farming practices, such as field burning and the use of gasoline-powered machinery, contribute significantly to the accumulation of greenhouse gases within the atmosphere. According to the Food and Agriculture Organization of United Nations (FAO), the livestock industry accounts for 18% of total emissions of greenhouse gases. The industrialization of the agricultural sector has increased the chemical burden on natural ecosystem which ultimately is becoming a threat to mankind. Pesticides, weedicides, chemical fertilizers cause a large number of negative health and environmental effects, their side effects can be important environmental health risk factor.

Agriculture is getting more and more difficult as a result of increasing population density of people and shrinking arable land. Hence, good agricultural methods should be used to deal with this circumstance and generate greater yield in a sensible manner that doesn't harm the environment in order to fulfill the demand for food. Every stage, from land preparation and tillage operations through post-harvest management, contributes to the production of high-quality crops and ensuring food safety.

Linkage between agriculture, food security, health

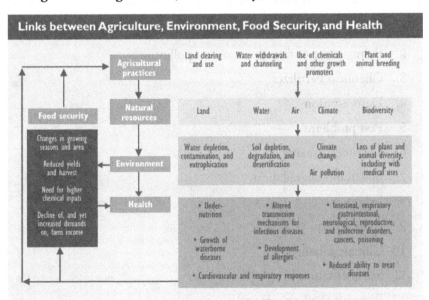

Figure 4: Schematic representation of the link between agriculture, environment, Food security and Health
[**Source:** Agriculture, Environment, and Health: Toward Sustainable Solutions by Rachel Nugent and Axel Drescher, (2006)]

Neglected Issues in Context of Agriculture

When it comes to environmental health, the most crucial factors from an agricultural standpoint are plant and soil health. Since they provide meat, milk, wool, and a variety of other products, ensuring the wellbeing of livestock is particularly crucial. An agronomist's top priority is to feed the rapidly growing population while increasing agricultural production, especially in low-income nations like India. Additionally, even though there is less land available for cultivation, more people are becoming dependent on it every day. In order to satisfy the increased rates of food consumption, farmers are working very hard to provide food to our growing population but in the meantime, wrong decisions are taken which affects nature. Modern agriculture has given its touch of innovation to the farmers but numerous unethical techniques are still being used in modern agriculture (crop production), which results in lower crop yields, lower output, and worse-quality final goods. Additionally, these unethical methods pollute the ecosystem, harm the soil, and upset the natural balance. Some of the negative effects which agriculture puts on the environment are:

1. Industrial & agricultural waste
2. Point source & non-point source pollution
3. Chemical Fertilizer
4. Deforestation
5. Pest problems
6. Livestock grazing
7. Soil/Land degradation
8. Irrigation
9. Biodiversity
10. Climate Change
11. Sedimentation
12. Riparian shading removal
13. Stream modification
14. Genetic engineering

1. **Clearance and use of land:** Land clearing and use for animal & crop production contributes greatly to soil problems like sodification (the accumulation of sodium), salinization, and depletion along with severe desertification. The UNEP or United Nations Environment Programme has estimated that active farmland suffers from soil loss of about 38%. These changes in the soil nature highly affect the production capacity with severe implications on food security. Clearance of the land, arable farming as well as animal production have been identified as primary factors for climate change since they tend to increase the concentrations of CO_2 and CH_4 in the atmosphere with notably respiratory and cardiovascular responses to changing temperature. These climate-change-induced temperature change increases the risk of occurrence of diarrhoea to 10% as per the estimates of The World Health Organization (WHO) in some areas. Moreover, the increased hurricane strength is also emerging out to be a serious issue associated with global warming. On other hand, various models predict that food production will be adversely affected by climate change. The practice is often responsible with the loss of medicinal plants & decline in the biodiversity of animals, plants as well as microbes for advancing the medical research. Clearing of natural vegetation often comes with the exposure of the top soil to be blown by the wind or washed away by the water along with increased flooding & sedimentation in water bodies.

2. **Controlled water use:** Agriculture is considered to be the largest water user for various purposes which not only alter or depletes the level but also contaminates as well as eutrophics it, which lead to poor human health. Unplanned or excessive irrigation led to increased breeding sites and habitat for vectors that transmit diseases like schistosomiasis (bilharzia), malaria, etc. Even the polluted water usage leads to the spread of viruses and parasites and consequently, diarrheal diseases, thus affecting the marine ecosystem.

3. **Use of agricultural chemicals:** To meet the global demand for feed and food, modern agriculture heavily relies on high inputs, especially chemicals like fertilizers and pesticides notably after the green revolution in 1960s. Though temporarily the problems of food and nutritional insecurity were resolved, but it brought us with variable issue of soil, water, air and plants including humans and animals (livestock) who

consume it. If we take an example of pesticides, we will be surprised to know that most of its residue stays in the environment or on food while less than 0.1% of pesticides actually reach the pests. Excessive use of different pesticides is also responsible for decline in the population of pollinating bees. Apart from these, in developing countries, the workers at pesticide plants or farmer using pesticides suffer from ill-health effects of pesticides every year. They suffer from chronic illnesses related to respiratory, neurological, intestinal, gastrointestinal, endocrine & reproductive disorders, as well as poisoning & cancers. These pesticides not only cause harm to humans but they also toxicate pollinators and wildlife. With drainage at fields, pesticides or chemical fertilizers passes onto the water bodies as well as groundwater, thus contaminating the resources. Pesticides which are used to control pests affect human's health by disrupting the immune & hormonal system, thereby increasing risk of cancer along with impaired brain development (Gilden *et al.* 2010). Excessive use of same insecticides for control of the vectors may sometimes lead in development of resistance such as in mosquitoes, which make them more difficult to control (Yadouleton *et al.*, 2009).

4. **Production of Genetically Modified Crops (GMC):** Crops where DNA has been modified using genetic engineering methods are known as GMC. Crops that are produced through genetic manipulation are not only detrimental to the health of humans but also to the environment as well. For the production of GM crops, there has been a huge requirement of agricultural input (chemicals), thus cultivation of GM crops have resulted to a huge increase in agricultural chemicals that are applied directly to crops. Along with that, main thing which puts the farmers in distress is the contamination of GM crops found in non-GM crop fields, as well as in soil and water systems. Moreover, we are also unaware about the future effects of consuming these modified crops with studies showing the carcinogenic effect of GMO (genetically modified organism) or GM crops Cotton, soybean Brinjal, Mustard and Potato (protein-rich potato) are some of the examples of Genetically modified crops).

5. **Global corporate agribusiness:** Agribusiness is the complete value chain in agriculture to create biological products to distributors and retailers that get products to end consumers, and represents all providers of value-added activities in agriculture. It links input providers, producers,

processors, and other services to consumers of crops, livestock, and other natural resources. Small and local farmers due to this global corporate agribusiness are really getting out of the business since they wouldn't be able to compete with lower prices of the commodity at the global market. This is the main reason why small scale farmers can't practice sustainable agriculture as they struggle with poverty, meeting the family's need.

6. **Integration of agriculture along with livestock in a standard agricultural system:** Modern agriculture has separated the rearing of animals and crop growing for increasing the efficiency, which existed in harmony before. If these enterprises are integrated, there would be no requirement of waste management with minimal requirement of fertilizer or commercial nutrient inputs.

Table 1: Ill practices followed in agriculture and their consequences in agriculture

S. No.	Bad practices that are common and practiced in many areas	Consequences of these bad practices
1.	Broadcasting of the seeds	Due to unequal space resources are either under exploited or over exploited. Improper crop geometry enhances the interplant competition for light, nutrients water etc., which affect the crop yield and quality.
2.	Conventional (Normal) or Excessive Tillage Operation	Deterioration of soil health More soil erosion Deterioration of soil physical fertility resulting yield loss
3.	Imbalance use of fertilizers (plant nutrients) i.e. excess or less in amount or use of poor quality compost or manure.	Less crop yield as well as poor crop quality Increase pest (weed, insect and disease) problem.

4.	Use of ground water for irrigation in excess amount as well as wrong method of irrigation.	Deterioration of soil health More water pollution due to contamination of heavy metals resulting in bad quality food / feed.
5.	Application of resources (fertilizer/ irrigation water) in improper time	One of the most crucial factors resulting in poor quality of crop and less crop yield, also soil contamination.
6.	Mono-cropping or Cultivation of only cereal crops	Soil fertility or soil health deterioration Increasing soil born insects-pests Developing resistance power in insect-pests, disease pathogens and weeds. More infestation of weeds, diseases and insect pests pushing farmers to use more plant protection chemicals causing more chances of residual toxicity.
7.	Indiscriminate (Unscientific) use of Plant Protection Chemicals	Poor quality of food / feed as there may be residual toxicity. Environmental pollution Deteriorate of soil health Disturbance of ecological balance Developing resistance power in insect-pests, disease pathogens and weeds etc.

Challenges Faced in the Linkage between Agriculture and Health Policies

We are very well aware that agricultural and health sectors have surprisingly failed to work in coordination and are given separate divisions within institutional organizations consisting of different function & worldviews. Improving food security and encountering malnutrition are though interdependent, but still we lack in coordinating these two communities closer together where the scarce resources could be utilized more effectively.

» **Institutional divisions:** A relatively rational ordering of tasks by the government tasks is out casted due to different sectoral government organization as we see in a different health and agriculture ministry along with their respective institutions or associations. There is always a reflection of self-contained nature with non-duplicative mandates in each and every sector. Though, we have noticed that the organization have individually performed best in overcoming the challenges regarding the development they face but still a competitive mindset is seen when there is a context of budget allocations. Due to this competition, collaborative efforts by the different sectors become too difficult. The institutions always carry forward a work based on their personal term and always verify that such work doesn't result in the net loss of resources for them. If so, then they are unwilling to share resources even when cross-sectoral approaches are optimal. We can understand it by an example like if the Ministry of Agriculture writes some proposals that include some health related components, there seems a face of sorrow in Health Ministry as if agriculture is trying to take resources that should be theirs. There is always a tendency of less collaboration and chances of conflicts occurs even in the resource constraint areas over allocations of the resources rather than increasing collaborative activity and maximize the use of what is available.

» **Differing functions:** Though there is a need of cross sectoral approach, but still we need to know that there is a concrete difference between the functioning of the two sectors to the wellbeing of the society. Agriculture can be considered as a productive activity where there is creation of economic value along with sustaining the livelihoods of the people. In contrary, the health sector is not a directly productive sector, but contributes majorly for the creation of a healthy households & society. So if we take an example that if the government works with the objective of fostering the economy, particularly for the underdeveloped agrarian societies, then agriculture will play a distinct role in development strategies for mitigating the conflicts. On other hand, if broad human development is the primary objective of government action plan, then the health sector is prioritised & agriculture plays a minor role.

» **Selective worldviews:** Both the health & agriculture sector have certain ideology and worldviews where they address and prioritize certain features while viewing the rest of the world and beyond their areas of expertise as irrelevant. The primary objective with which an agriculturalist work is maximizing the production and productivity while the health professionals works with a motive of providing good health care facilities and preventing outbreak of any epidemics. Apart from this, different institutional backgrounds & training paths hinders the development of any common focus and both the sectors have different parameters of assessment of its own success which sometimes may not successfully correlate with each other.

Chapter - 12

Agriculture and Health Linkages in the MDGs

The Millennium Development Goals or MDGs (2015) refers to the first global policy framework for reduction of the poverty adopted at a level where opportunities for collaborative approaches can be enforced and overcoming of the sectoral divides can be successfully performed. The MDGs are therefore considered as a useful framework for the identification of those areas where joint policy formulation between health & agriculture can bring about significant benefit in reduction of the poverty.

Table 2: Agriculture and health linkages in the MDGs

Sl. No.	Millennium Development Goals	Synergies between health and agriculture
1st Goal	Eradication of extreme hunger and poverty	» Poverty can be greatly linked with better health, which in turn also sustains the resource base for agriculture
		» Creation of a healthy and secured agricultural livelihood depends much on the health of its members. If the children & adults are sick and need to care for themselves, then the whole livelihood becomes less productive

		» High health costs are generated if there are occurrences of ill health conditions in relation to agricultural production, thus lowering the income of the rural and semi-urban people
		» Different impacts have been observed on nutrition, health, and well-being of an individual with the adoption of different agricultural production systems
		» Profitable agricultural production can improve the access to better health products as well as facilities to the needy households
		» People's nutritional status as well as nutrient absorption is sometimes affected by agriculture-associated infections
2nd Goal	Universal primary education achievement	» Areas having a healthy rural communities have always witnessed a less participation of children in household chores as well as agricultural production which in turn leads to reduced absenteeism
3rd Goal	Empowerment of women & promoting gender equality	» There must be equal participation of both the genders in carrying out the agricultural work, which also serves the benefit of focussing on the vulnerability related to gender & reduction of health risks related to specific agricultural tasks
4th Goal	Reduction in child mortality rates	» Less occurrences of illness associated with agro-ecosystems, improved environmental management along with better nutrition lead to healthy mental and physical growth of children, thus having a sharp decline in under-five mortality & childhood illness.

5th Goal	Improvement in the maternal health	»	The increased chances of having safe & healthy pregnancy are observed with better maternal nutrition & health along with the ability to engage in agricultural activities
		»	Pregnant working women can be benefitted with the occupational health policies in agriculture.
6th Goal	Combatting diseases like Malaria HIV/ AIDS, etc.	»	Adoption of integrated pest management and integrated vector management along with correction in certain environmental management practices in agriculture contribute a lot in reducing the risk of malaria transmission
		»	Rural communities with a proper livelihood from agriculture should be educated and made aware about the risk of associated sexually transmitted diseases with the risky sexual behaviour, who perform it as a source of additional income and thus combatting STDs like HIV/AIDS
		»	The farmer's potentiality in performing the agricultural production is enhanced by the less pressure by infections in the immune system.
7th Goal	Attainment of environmental sustainability	»	Balanced and sustainable use of water resources for domestic and agricultural purpose supports communities on a healthier scale

		»	Problem of water scarcity in agriculture along with protection of the health can be encountered with the usage of excreta, wastewater and grey water as valuable resources
		»	Chemical inputs should be judiciously used in agriculture keeping in mind the adverse effects of contamination of surface and groundwater, thus promoting a safer health.
8th Goal	Establishment of a global partnership for development	»	Enhancement in the health potential of agricultural development projects can be performed by various impact assessment procedures by national, bilateral and multilateral agencies.

Chapter - 13

Crop Management and 'One Health'

There is a huge risk to health at the global level. The entire health system of the mankind possesses an enormous burden due to some serious life-threatening and life limiting illnesses. There is a need for a fundamental reshaping of human relationship with nature due to alarming increase in the global population. The selfish nature of the people has now become the root cause of the problem for the deterioration these days. Importance is given only to the human health. People generally have a misconception that with the maintenance of good human health, there could be minimization of all health-related issues. But this goal cannot be achieved until and unless prioritization is focused equally on the health of other living components of the globe which includes the reared livestock of the farm, etc. Along with the health of agricultural produce, soil on which produce is grown must be given greater importance.

Agricultural practices that are generally associated with crop production which includes cultivation, soil management, and row cropping are known as agronomic practices. These practices serve as a vital part of the entire farming systems. They are incorporated by farmers to manage crops, enhance water usage, improve soil quality & improve the overall environment. Better fertilizer management lies as prime way of improving agricultural practices. There has been decrement in the costs of input involved in producing farm products with the proper usage of agronomic practices. Consequently, quantity of the yield along with the quality showed an increment significantly. The quality of land can be maintained with proper use of fertilizer and decreasing water usage by crops.

Some agronomic practices are:-

» **Preparation of land (soil):** The soil needs to be prepared prior to planting to kill the weeds in the seedbed that would compete with the crop for light, nutrients and water with the chemical "burn-down" or any form of tillage. Residue left over the surface slows down the runoff flow which has the potential to displace and carry away soil particles.

» **Method / process of sowing:** Before the sowing operations to be carried out, seeds should be placed in medium hot water for a period of at least 24 to 48 hours. Cleaning of the seed is done especially with fruit crops as they are more prone to attack from insects or pests. Seed are usually cleaned by rubbings with a cloth/paper sometimes assisted with a seed washing. Washing of the seed is generally done by submerging cleansed seeds in 50°C water for 20 minutes. Seed washing with high temperature or hot water becomes more vital for tropical fruits which are easily infected such as litchi (lychee or leech) and rambutan (*Nephelium lappaceum*).

» **Manures and Fertilizers addition:** For proper growth and development, nutrients/ food/ elements are required by plants which are absorbed through soil. The sources of supplying nutrient are manures and fertilizers. It is considered as an important factor that helps in maintaining the soil fertility and increasing the crop yield.

Manure: Well decomposed refuse from the stable and barn yards including both straw or animal excreta and other litters are termed as manure. Its main objective lies in supplying the plant nutrients to the soil making it productive as well as promoting plant growth.

Fertilizers: Industrially manufactured chemicals which contain the essential plant nutrients, essential for growth of plants are termed as fertilizers.

» **Irrigation:** Artificial application of water to the soil or land is termed as irrigation. Water is supplied to dry land by means of ditches, ponds, canals etc. It is the. Its main usefulness lies in maintenance of landscapes, assisted in the growth of agricultural crops along with re-vegetation of disturbed soils in dry areas during inadequate rainfall. Other uses include such as protecting plants against frost, etc.

» **Weed management**: Weeds are those plants growing where they are not wanted and always in competition with cultivated plants. They compete with productive pasture or crops, thus converting the productive land into unusable scrub. Weeds are also often distasteful, produce burrs, thorns or other damaging body parts and even poisonous and responsible for contamination of the harvests.

» **Other crop pests management**: Nearly 30% crop loss occurs due to pest attack. Pests can be categorized under various types starting from sucking type, piercing type, etc. For getting optimum yield, agronomic practices must be monitored for proper disease and pest control.

» **Harvesting**: Getting the crop out of the field after attainment of its maturity is termed as harvesting which is then transported to market. Except hay, which is cut several times, most of the crops are harvested in the fall. Small fruits and other food crops are typically harvested by hand except for few cases while field crops are harvested by machine. Most commonly used equipment for harvesting are: Tractors, Combines, Grain Carts, Harvesters, Cotton Harvesters, Balers, etc.

» **Grain storage**: Prevention of loss of quality of the grain due to moisture, weather, and wind along with birds, rodents, insects, and microorganisms is the main purpose of any grain storage. Grain storage can be done in pole buildings, stud framed shops or garages and empty barns. Protection must be guaranteed against insect damage for seed stored for more than six weeks. Grain containing high moisture causes a rise in temperatures which lead to development of mould. Rise in the temperatures allow the multiplication of insects and reduce the efficacy of grain protectants. So, seed should only be stored only when 'dry'.

Impact of agronomic practices on 'One Health':

Agronomic practices can have both positive and negative impact on one health that is environmental, human, plant and animal health depending on the specific practices employed and the context in which they are used.

Negative impact:

Agronomic practices can have negative impacts on human health, also known as One Health. Here are some examples of negative impacts of agronomic practices on One Health:

Disease Transmission: Agronomic practices that involve the raising and slaughtering of animals can increase the risk of zoonotic disease transmission, which occurs when pathogens that infect animals are transmitted to humans. Zoonotic diseases such as avian influenza, swine flu, and salmonellosis can have serious health consequences for humans, including respiratory illness, gastrointestinal illness, and even death.

Antibiotic Resistance: Antibiotic use in agriculture can contribute to the development of antibiotic-resistant bacteria, which can have serious health consequences for humans. Antibiotic-resistant infections can be difficult to treat and can lead to increased morbidity and mortality. Overuse of antibiotics in agriculture can contribute to the development of antibiotic-resistant bacteria by promoting the selection of resistant strains.

Deforestation and Loss of Biodiversity: Agronomic practices such as clearing land for crops and livestock can contribute to deforestation and loss of biodiversity. Deforestation can lead to habitat loss for animals and plants, which can have negative impacts on ecosystem services such as pollination and nutrient cycling. Loss of biodiversity can also increase the risk of zoonotic disease transmission, as intact ecosystems can help regulate the spread of disease.

Positive impact:

Agronomic practices can also have positive impacts on human health, also known as One Health. Here are some examples:

Access to Nutritious Food: Agronomic practices can increase access to nutritious food, which is essential for good health. This includes the production of grains, vegetables & fruits, which are important sources of minerals, vitamins and fiber.

Physical Activity: Some agronomic practices, such as gardening and farming, can provide opportunities for physical activity. This can help reduce the risk of chronic diseases such as heart disease, obesity, diabetes, etc.

Sustainable Agriculture: Sustainable agricultural practices, such as agroforestry & organic farming can reduce the use of harmful chemicals and protect the environment. This in return can reduce the risk of exposure to toxic substances and help improve water and air quality.

Livelihoods and Food Security: Agronomic practices can provide livelihoods for rural communities and help ensure food security. This can improve overall wellbeing and health and help reduce poverty.

Biodiversity Conservation: Agronomic practices can help conserve biodiversity by preserving natural habitats and protecting endangered species. This can have positive impacts on human health, as biodiversity is essential for maintaining ecosystem services such as clean air and water, and regulating the climate.

Some agronomic practices to achieve 'One Health'

Sustainable crop production: An interconnection between human health, animal health, and the environment is proposed via the concept of 'One Health'. Since food links the environment with the human health, agricultural production serves as a critical component of 'One Health'. Food serves as an important pathway for human exposure to potentially harmful agrochemicals as well as environmental microbes besides providing nutrients to humans. In addition, unintended adverse impacts on human health due to change in the environment can be brought with inappropriate agronomic practices. Therefore, protection of the environmental health along with improvement in agricultural production systems should be always interlinked without being viewed as isolated goals. One of the key goals of the 'One Health' concept is the utilization of full potential of micro biomes for better sustainable agronomic practices.

Use of organic protectants in preservation:

For the preservation of commodities, usage of organic and herbal protectants must be referred. These organic protectants repel pathogens and insects and even do not cause ill effects when consumed along with the produce.

Prevent surface runoff of inputs from fields:

An agronomist should try to reduce the surface runoff of chemicals from the field along with water. These chemicals when drop to water bodies, makes it polluted. These contaminants may even enter the body of fish which can cause harm to animals and humans feeding in it. Even when animals drink water from such water bodies, contaminants reach their system and causes hazard.

Good quality crop residue as feed for livestock:

The residues of the crop to be used as feed for the livestock should not be toxic. It should be free of weed seeds and other contaminants. It should not be too moist or too dry. Too moist will invite fungal pathogens to colonize it thus making it hazardous for animals to feed on it. An agronomist should ensure a good nutritional quality fodder is given to animals as feed.

Adoption of organic farming:

Organic farming is a solution towards sustainable agriculture. Organic farming should be practiced to get organic produce which are safe to consume and free of chemicals. Even organic farming is good for soil health as chemicals are excluded. so it cannot degrade soil health. However does organic farming alone can feed the ever rising population is a big question. So it is better to follow integrated farming practices which do not solely depend on chemicals.

Agricultural practices can have both positive and negative impacts on One Health, and it is important to consider these impacts when designing and implementing agricultural policies and practices. By promoting sustainable agriculture and adopting practices that are compatible with One Health principles, we can improve the health and well-being of people, animals, and the environment. Agronomic practices play a crucial role in promoting sustainable agriculture and ensuring food security. By adopting soil conservation practices and integrated pest management, farmers can improve crop yields, protect the environment, and promote human and animal health. As such, it is essential to prioritize the implementation of agronomic practices that are both sustainable and promote 'One Health'.

Integrated Farming System (IFS) and "One Health"

Integrated Farming System (IFS) integrates natural resources in an effective and efficient manner to secure sustainable production of better quality food and other products through eco-friendly technologies, sustain farm income, eliminates or reduces environmental pollutions generated by conventional agriculture; it also help to achieve maximum replacement of off-farm inputs. The main purpose of integrated farming system is to bring self-sufficiency in farmers' requirements of food and cash (money). IFS increases income and employment opportunities as well as recycles farm by-products and crop residues / farm wastes and thus, increase resource use efficiency through efficient management of resources. .

Goals of Integrated farming system

i.   Maximization of yield through the combination of various enterprises to provide steady & stable income at higher levels.

ii. Achievement of agro-ecological equilibrium and rejuvenation of system's productivity.

iii. Aims at controlling the build-up of noxious weeds, insect-pests and diseases and maintaining them at low level of intensity with help of natural cropping system management.

iv. Reduction in harmful agro-chemicals and pesticides and use of chemical fertilizers to provide pollution free, clean and healthy environment to the society at large scales.

v.   Maximizing nutrient use efficiency and minimizing nutrient loss.

Role of IFS in 'One Health'

In IFS, waste of one component becomes an input for the other. The inputs are recycling continuously through a system. Thus the dependency on external inputs is deduced thereby protecting the agro-ecosystem as a whole. The harmful effects of chemical input residue on crops is deduced and also no chemical residue is taken up by human being and animal. Due to less chemical input pressure, the volatilization loss from this inputs is not contributing much to the greenhouse gas content in the atmosphere thereby protecting the environment .Thus it is playing a pioneer role in promoting "One Health". This in-turn raises the overall profitability by complementing main allied enterprises with each other. IFS on large scale focuses on the use of on-farm inputs, reducing the chemical fertilizers and pesticides usage and thus establishing a linkage between the human health as well as environment. Due to low usage of chemicals, a chance of eutrophication as well as biological oxygen demand is maintained. The practices involved in IFS also lead to better weed control and supplementation of weeds like *Hydrilla* into the fields with the purpose of nutrient enrichment. IFS also serves as an efficient crop residue management tool where there is least occurrences of stubble burning along with organic matter addition in the field.

Chapter - 14

Role of Agronomist

In the rapid stage of changing climate and uncertain energy sources, scientists are working continuously to provide us with safe food and water using the natural resources left in sustainable way. The most critical problems agriculture and natural resources had faced in 2015 is providing clean and abundant water, responding to climatic variability, improving human health, developing renewable energy, strengthening food safety and establishing sustainable food production. In spite of area under cultivation being static, population pressure is increasing. To solve this problem, on the same piece of land in a year more number of crops has to be grown. So, farmers adopted conventional methods of farming as well as inorganic modes to feed the huge population. Few best management practices which are affordable and practical approach to conserve farm's soil and water resources without sacrificing productivity are:

» **Organic amendments:** Soil's physical properties, such as permeability, water infiltration, drainage, water retention, aeration are improved due to high soil organic matter content. Ultimately it provides a better growing environment for crop roots. Adding suitable organic amendments is one of the most effective ways to build and maintain levels of our soil's organic matter.

» **Choice of variety:** The variety to be chosen before seed sowing must be a high yielding variety producing more yield along with both biotic and abiotic stress resistance (biotic and abiotic stress free), which ultimately avoid the use of harmful plant protection chemicals.

» **Crop Replacement:** We all are concerned about choosing an appropriate variety of a crop for a particular location. Whenever there is emergence of abiotic or biotic stress, we look in alteration or adoption of variety be it resistance or tolerance. But in recent years, there has been turnover from variety replacement to crop replacement. Let us take an example of the dryland areas of India. These areas generally grow crops that require less water. So, instead of traditional crops grown in these areas, we can switch over to millets, tuber crops or legumes. *Tubers, pulses and millets are important for the livelihoods and nutrition of poor farmers, especially in fragile or resource poor regions.*

The millets, also known as 'nutri-cereals', are one of the oldest foods known to humanity, not only require less water but also low agricultural inputs. Sorghum, pearl millet, finger millet are known as major millets; foxtail, little, kodo, proso and barnyard millet are known as minor millets. They have the potential to generate farmer's income along with food and nutritional security across the world. These millets feed more than a billion people across Africa and Asia. Similarly pulses are known as climate-smart crops. Pulses and millets which are an important component of food and nutritional security of the poor, need less water and produce assured and high yield. Pulses support a sustainable and diverse agriculture system, use minimal water to grow, have a low carbon footprint when used in cropping rotations, and are a nutrient-dense source of protein and contribute to food security and human nutrition across the globe. Tuber crops adapt to a wide range of agro-climatic conditions and give good performance even under marginal growing conditions.

» **Conservational tillage or reduced tillage or minimal tillage:** Tillage results in physical soil disturbances which lead to base or compacted soil leading to destruction to soil microbes and creation of a hostile environment to live for them. Conservation tillage reduces erosion by protecting the soil surface and allowing water to infiltrate instead of running off. Mulch tillage is an example of conservation tillage where crop residues are left on the surface, and subsurface tillage leaves them relatively undisturbed. Mulch tillage systems disturb soils less than conventional tillage systems. The remaining residue cover provides soil protection from wind and water erosion, increases the amount of organic matter in the soil, and conserves soil water by checking evaporation at the soil surface. It may also be helpful in reduction of weed population.

» **Diversify soil biota with plant diversity:** Crop diversification enhances soil organic carbon (SOC) storage, soil biodiversity and ecosystem functions. Crop diversity through proper / scientific crop rotation increases the quantity, quality and chemical diversity of plant-derived carbon inputs to soil, as a result growth and diversity of soil microbial communities is encouraged. A diversity of plant carbohydrates is required to support the diversity of soil microorganism in the soil which can be achieved by crop rotation.

» **Crop rotations:** Growing of different types of crops in recurrent succession on the same field / land is known as crop rotation. In several parts of the world, monoculture or single cropping is favoured by farmers but it is the role of an agronomist to open up the view of diversified farming. Growing of different crops in sequence in the same plot to improve soil health, optimize soil nutrients, and counteract the pressure of pests and weeds. It not only brings economic benefits but also protect crops from pests and pathogens. For increasing the soil fertility, introduction of legume crop can be viable alternative in crop rotation. As the legumes fix a high amount of nutrients, so lesser amounts of nutrients are needed as input thus, yielding a good quality produce and maintaining a better soil health. During the decomposition of organic matter or its break down, N is released either into soil microbial biomass or into soil where it can be easily utilized/ used by the plants. During this process, some portion of N is released into the soil for succeeding crop known as 'Legume Effect'). Changing the crop grown from year to year in a particular field will provide a variety of root systems and different types of crop residues. This leads to improvement in soil structure and maintenance or improvement in soil organic matter levels.

» **Harvesting:** To get optimum yield, harvesting of the crop at right time and method will ensure a better quality of the produce. Timely harvesting ensures good grain / seed / food quality and high market value, moreover, yield loss can be minimized by harvesting on time.

There will be less chance of pest and insect infections and produce of high nutritional quality. It is an important role of agronomist to guide the farmers for getting better crop yields and optimum returns. Quality of the produce becomes very important as it is the main thing moving on higher trophic level.

» **Adoption of line sowing of crops in place of broadcasting:** Practices like broadcasting not only increases the seed rate per area but also in turn cost of cultivation. Even the crops are more prone to lodging leading to less crop yield. Recent research clarifies broadcasting to be less scientific method of sowing. Insect, pest and disease attack is more due to overcrowding as well as irregular spacing and consequently spraying of more chemicals showing a negative impact on environment and soil. Agronomist should help the farmers in adoption of such practices.

» **Stubble Burning:** Lack of education among the farmers result in burning of the stubbles of the previous crops. It is followed in few states of the country. It is a serious issue and is considered a major event in causing 'Smog' in Delhi which degrades the air quality and causes respiratory disorders in humans. As a result, scientist discovered 'Pusa Decomposer' which decomposed crop residue, including paddy straw and turns it into manure in about 25 days

» **Land retirement:** Land retirement is a practice that takes agricultural land out of cultivation due to poor soil condition that means land is kept fallow. As the amount of soil eroded is too great to economically return to the upper slope, some upland portions of fields are not suitable for remediation. There is a greatest risk of emissions of GHG like methane and nitrous oxides from the marginal lands which are too heavy or too wet to crops. Habitat conditions could be improved by retirement to shrubs or natural grassland species and trees. It also has an advantage of rehabilitate the soil health and provide long-term cropping options for future generations.

» **Crop Residue management:** Harvesting of various crops generates large volume of residues both on and off farm. Crop residues may be used for mushroom cultivation, soil mulching, for compost preparation, as bedding material for animals, livestock feed, bio-gas generation, thatching for rural homes, biomass energy production, fuel for domestic and industrial use, etc. Effective residue management can lead to better soil cover. Effective cover begins at harvest. Complete mulch effect will be provided with uniform cover of crop residue and manage will be easier when planting the next crop. The problem of 'on-farm' burning of crop residues is increasing in recent years due to shortage of agricultural labour, high cost of removing the crop residue from the field and mechanized harvesting of crops.

» **Balanced use of synthetic fertilizers:** Synthetic fertilizers have a short term positive effect on productivity but a long term effect on soil or environment. Their use efficiency is also less. Agricultural chemicals like commercial fertilizers, pesticides, weedicides, fungicides and other plant protecting chemicals should be used at a minimal rate. IPM, INM and IWM helps in maintenance of soil fertility and protection of crops for sustaining crop yield production through optimization of benefits from all possible sources of plant nutrient in an intEGRATED MANNER

» **Cover crops:** Cover crops are grown for the protection and enrichment of the soil. They are proven to be beneficial to crops as well as soils. One key soil health benefit is that they provide cover when crop residue or other living crops cannot. For protection of the soil from wind and water erosion, a solid stand of a drilled or broadcast cover crop should be used. Green manure crop may be used as cover crop.

» **Erosion control structures:** Construction of a water and sediment control basin (WASCoB) to control the rill erosion. WASCoBs when complemented with other soil conservation best management practices (BMPs) such as crop rotation, mulching, cover crops and residue management along with other erosion control structures are most effective in control.

» **Usage of more surface water rather than groundwater:** An agronomist should use more surface water instead of ground water for various uses like irrigating crops etc. Excess of groundwater usage increases the problems of arsenic and other heavy metals in plants and soil.

» **Rainwater harvesting:** Collection and storage of rain water for reuse is known as rainwater harvesting and it should be done for irrigating crops. Rainwater serves as an excellent source of sulphur which makes the plant resistant against fungal diseases as well as sucking insects. Harvesting of rainwater reduces runoff, soil erosion, flooding, and pollution of surface water with fertilizers, pesticides, metals and other sediments.

» **Using weed plants in positive prospect:** Aquatic weeds like water hyacinth and hydrilla can remove heavy metals from soil or aquatic system and serve for bioremediation of lead mercury and cadmium. Agronomist must educate the farmers to use weeds to reduce contamination

» **Not keeping the soil fallow:** Fallow or bare land are more prone to erosion and also problem of weeds may be seen. Soil must be covered with crop residues or mulch crops. It is always advisable to keep the land cultivated with crops in rotation.

» **Use of Nano-Technology:** Now–a–days, nanotechnology emerges out to be promising field supplementing us with different nano-devices and nano-materials having a unique role in agriculture. Some examples are:-

1. **Nano-biosensors:** It is capable of detecting nutrient status as well as moisture content in the soil and also applicable for site specific nutrient and water management

2. **Nano-fertilizers:** It has proven to be efficient in nutrient management. It has greater role in enhancing the production of crops, reducing the cost of fertilizer application and also reduces the pollution hazard. These fertilizers contain nano-structured formulation which controls the release speed of nutrients for matching the uptake patterns of the crop.

3. **Nano-herbicides:** It is effective in the selective control of weed in crop field

4. **Nano-nutrient particles:** It increases seed vigour.

5. **Nano-pesticides:** It has proven to be efficient in pest management etc.

»   **Bio-fortification:** "Biological fortification" or bio-fortification" is defined as process of producing nutritionally enhanced food crops that are developed using conventional plant breeding, modern biotechnological techniques along with agronomic practices. There are many success stories of bio-fortification out of which two i.e. lysine and tryptophan rich quality protein maize along with Vitamin A rich rice, orange and sweet potato prove to be promising. Even the leading organisations like World Health Organization (WHO) and the Consultative Group on International Agricultural Research (CGIAR) as a part of their main goals includes the development of high-yielding nutritionally enhanced bio fortified crops.

Chapter - 15

Pulse Crops and 'One Health'

Pulses are leguminous crops, including green gram, black gram, pigeon pea, pea, lentil, chickpea, etc. which produce edible seeds that grow within a pod. The word "Pulse" is derived from the Latin words '*puls*' or '*pultis*' meaning 'porridge', a preparation, a dish, or 'thick soup'.

Our dietary dependence on pulses is enormous because majority of the Indian are vegetarian and pulses are the most economic source of protein for them. Pulses provide protein and fibre, as well as vitamins and minerals, such as iron, zinc, magnesium, etc. However, pulses in India are traditionally considered to be an **orphan crop,** mainly grown under marginal land or stressed situation as rainfed crop.

An expert group on pulses reported that **pulses are the richest and the most economical source of protein** (Sengupta, 2022).

Cultivation of different pulse crops (or at least one pulse crop in sequence or crop rotation) is essential in our cropping system as because pulses play an important role for a number of reasons.

» Pulses are low in fat and rich in soluble fibre, which can lower cholesterol and help in the control of blood sugar.

» They are packed with nutrients and have high protein content, thus, an ideal source of protein particularly for the people where meat and dairy products are not physically or economically accessible.

» Pulses are a source of plant-based proteins, amino acids and other nutrients in human diets. Pulses are recommended for the sick people and management of diseases like diabetes and heart problems.

» Pulses may help to combat obesity.

For farmers, pulses are very important crops because:

1. Pulse crops improve soil fertility;

2. Pulses improve soil productivity;

3. Water requirement is comparatively low and can be grown as rainfed crops;

4. Farmers can sell the produce (seed / grain / fodder) at a higher price; and

5. Their family members can consume the produce.

Thus, pulse crops can help the farmers to maintain household food and nutrition security and give them economic stability.

Pulse crop is one of the most secured crops farmers should cultivate in their crop sequences. Pulses can better withstand climate change thus reducing risk for the small or marginal farmers. As a multi-use crop (food, fodder, fuel, green manure, etc.) pulse can improve livelihoods of farmers, particularly for female farmers who are a major part of the labour force in our farming.

Pulses play an important role for sustainability in many ways. They are an important component of crop rotations, they require fewer amounts of fertilizer inputs than other crops and they are a low carbon source of protein. Pulses are considered as most sustainable crops, because they add atmospheric nitrogen (N_2) into the soil; they are helpful for checking the soil erosion as they have more leafy growth and close or dense covering; pulses play important role in crop rotation, mixed and intercropping; and add organic matter into the soil. The average rate of N-fixation of pulses is about 1.0 kg/ha/day within a cropping season, which generally considered as potential N-fixing ability in a given environment (Sengupta, 2022). Pulse crops, on an average, can fix 30-150 kg N/ha depending upon rhizobial strain and population, host crop and varieties, soil properties, agronomic management practices and environmental conditions, and out of this nearly 65-70% of the nitrogen fixed becomes available to succeeding crop growing in the next season. Thus, considerable amount of N-fertilizer (synthetic fertilizer) input can be saved in succeeding crop. Moreover, biological nitrogen fixation provides a natural slow-release form of nitrogen supply that usually matches crop needs.

Approximate amount of N- added into soil by different pulse crops

Crop	Amount of N- fixed (kg/ha)
Black gram (*Urd* / *Urad* bean)	50-60
Pigeon pea (*Arhar*)	68-200
Green gram (*Moong*)	50-55
Lentil (*Masur* / *Masuri*)	35-100
Chickpea (Bengal gram)	26-63
Field pea (*Matar*)	46-50
Cowpea	53-85

[Source: Gill *et al.* (2009) and Brahmaprakash *et al.* (2004)]

Pulse crops maintain sustainable ecology. Introduction of pulse crops in cropping sequences can lower the average carbon footprint by 24 to 37% (ICRISAT, 2016). Pulses can reduce use of synthetic (chemical) fertilizers by fixing atmospheric nitrogen. Thus, on an average pulses reduce non-renewable energy in the entire crop rotation by 22-24% in the entire crop rotation.

Pulses provide / add a considerable amount of organic matter to soils.

Pulse crops are well-suited to low moisture conditions. Scientists found that chickpea and lentil produced good yields even when water was limited. Under severe drought conditions, when some crops failed and did not give any appreciable yields, chickpea and lentil were able to maintain some yields.

Pulses are also a good protein source with a low footprint, in both carbon and water. Research work revealed that the water footprints to produce 1 kg of beef, pork and chicken are 43, 18, 11 times higher than the water footprint of pulses (Sengupta, 2022).

Pulses have a lower carbon footprint in production than most animal sources of protein.

Pulses contribute to reduce the emission of greenhouse gases (GHG), as they release 5–7 times less GHG per unit area compared with other crops. One study showed that 1 kg of legume only emits 0.5 kg in CO_2 equivalent, whereas 1kg of beef produces 9.5 kg in CO_2 equivalent.

Pulses have a very low Global Warming Potential (GWP) values. They are drought tolerant, thus use less water. Low water use results in low energy use. Introduction of pulse crop in cropping system can increase the water use efficiency of the entire cropping system (ICRISAT 2016).

Pulse-based cropping systems improve several aspects of soil fertility, such as SOC (soil organic carbon) and humus content, N and P availability. Pulses can supply huge amount of biomass, organic C, and N. The plants are carbon rich, thus when added / incorporated into soil as crop residues or green manures, enhance C-sequestration in the soil through rapid decomposition. The crop residues decomposed easily and quickly due to low C / N ratio as a result soil fertility is improved (Sengupta, 2022).

Pre-crop benefits of pulses (known as 'Legume Effect'):

1. Nitrogen effect (addition of soil N by biological nitrogen fixation), and

2. Break crop effect (improvements of soil organic matter and structure, phosphorus mobilization, soil water retention and availability, and reduced pressure from diseases and weeds).

Pulse crops produce a number of different compounds that feed soil microbes and help to maintain soil health, thus, they have a significant impact on soil biology, increasing soil microbial population and activity even after the harvest. Crops' performance is better in soils that are more "alive" with a diverse array of soil organisms.

Phosphorus (P) is an essential macronutrient for plant growth and metabolism. However, phosphorus nutrition is a major limiting factor for crop production in many soils due to relatively low P availability because phosphorus can be readily adsorbed or fixed in soil. Pulses can mobilize fixed forms of soil P (phosphorus) through the secretion of organic acids. Pulses enhance rhizosphere (the region / layer of soil in the vicinity of plant roots) chemical processes more than cereal crops to mobilize sparingly soluble soil P by rhizosphere acidification and enhanced exudation of carboxylates and phosphatases (Sengupta, 2022).

Rose *et al.*, (2010) reported that pulses have the ability to free soil-bound phosphorous, thus, play an important role in the nutrition of plants.

Pulses exude greater amounts and different types of amino acids (than non-legumes) and the plant residues left after harvesting have a different biochemical composition (e.g. better C : N ratio) than other crop residues.

Intercropping has a higher soil carbon sequestration potential than monocrop systems. Pulses (Legumes) play a pivotal role in many intercropping systems. Among the top ten (10) most frequently used intercrop species seven (7) are pulses (legumes). By using pulses for intercropping and cover crops, farmers can maintain farm and soil biodiversity.

Despite their important role in improving the sustainability of cropping systems and mitigation of the ill-effects of climate change, pulses have not received the same attention compared to cereal or other non-legume crops.

Pulses are high in fibre and have a low glycaemic index (GI - it is a rating system for foods containing carbohydrates and indicates how the food affects blood sugar level of our body), making them particularly helpful to people with diabetes by assisting in maintaining healthy blood glucose and insulin levels (Mudryj *et al.*, 2014). Pulses have further beneficial effects on health because they possess antioxidant and anti-carcinogenic properties so they have significant anti-cancer effects. Pulse consumption also improves serum lipid profiles and positively affects several other cardiovascular disease risk factors, such as blood pressure, platelet activity, and inflammation (Mudryj *et al.*, 2014).

Climate smart sustainable farming is possible through pulse cultivation thus, inclusion of one pulse crop in crop rotation is very important. Even in agro-forestry systems or in watershed areas, inclusion of pulse crops like pigeon pea (*Arhar / Tur*), may support adaptation through diversification of the income source, increased productivity, increased resilience to climate extremes, and decreased runoff and soil erosion. Cultivation of pulse crop in dry or rainfed areas is found beneficial since the crop can add organic matter and nitrogen, check soil erosion and support the population and activity of soil microorganisms. Pulses are climate smart as they simultaneously adapt to climate change and contribute towards mitigating its ill effects. Our food demand is increasing at a fast rate due to enormous population growth that has posed a challenge to meeting the requirements of nutritionally balanced diets. Pulses could play a major role in the human diet to combat these challenges and provide nutritional and physiological benefits. Pulses may play an important role in providing valuable products for animal feeding and thus indirectly contribute to food security. Pulses and their by-products are extensively used as animal feed. Pulses play a crucial role in sustainable development due to their nutritional, environmental and economic values. Malnutrition is a predominant problem in our country. Hence, nutrition oriented sustainable agricultural development is of utmost necessity in the present context. **Therefore, pulse crops should be an integral part of sustainable agro-ecosystems.** Pulse crops are one of the most sustainable crops a farmer can grow. **Farmers must include pulse crop in their cropping system to enrich the soil with nutrients, create better growing conditions, promote biodiversity and reduce the need for synthetic fertilizers.** Growing pulse crops in rotation with other crops enables the soil environment to support the large, diverse populations of soil organisms. The ability of pulses to feed the soil different compounds has the effect of increasing the number and diversity of soil microbes (Sengupta and Biswas, 2017).

Conservation Agriculture

In conservation agriculture permanent vegetative cover (or mulch) is maintained on the soil surface, there is minimal soil disturbance or movement (by taking zero / reduced tillage practice) and diversified crops cultivation through proper crop rotation or intercropping. Conservation agriculture conserves biodiversity, natural resources, and labour and increases the availability of soil water, reduces different crop stress, and maintains or builds up soil health in the longer term. The core principles of sustainable agriculture and conservation agriculture are more or less same. The main aims of conservation agriculture is to protect the soil from degradation or erosion and to increase the crop yields. In conservation agriculture inputs are efficiently used; basically it is a resource-conserving agricultural production system.

Chapter - 16

Sustainable Agriculture

Sustainable agriculture tries to find a desirable balance between the need for food production and the preservation of the ecological system within the environment. Sustainable agriculture is a type of agriculture that focuses on producing long-term crops and livestock while having minimal adverse effects on the environment. It is a type of farming where we can expect maximum production of crops or agricultural yields to meet the needs of present generation without endangering the resource base of future generation. By adopting sustainable practices, farmers may reduce their reliance on non-renewable energy, reduce chemical use and save scarce resources.

Goals of sustainable agriculture:

» Production of sufficient food, feed, fibre, and fuel to meet the needs of a sharply rising population.

» Protection of the environment or environmental health, and expansion of the natural resources supply.

» Sustainment of the economic viability of agriculture systems (i.e., economic profitability, and social and economic equity.

Fundamental differences between sustainable and conventional farming

Sustainable Agriculture (SA)	Conventional farming (CF)
Synthetic fertilizers and synthetic pesticides are rarely used or not permitted	Synthetic fertilizers and synthetic pesticides are allowed and commonly used
Genetically modified organisms (GMOs) are rarely recommended or not allowed	GMOs can be used
Soils have better health and higher water holding capacity than CF	Soil health is gradually deteriorating and have less water holding capacity than SA
SA has larger floral and faunal biodiversity than CF (complex crop pattern)	CF has smaller biodiversity than SA (simple crop pattern)
The agricultural landscape is characterized by heterogeneity (multicultural system)	The agricultural landscape is characterized by homogeneity (monocultural system)
The use of non-renewable resources is minimized by recycling plant and animal waste into the soils (on-farm inputs)	Depends largely on non-renewable resources (off-farm inputs)
Crop protection depends mainly on natural processes such as soil fertility, crop cycle, and biodiversity (more preventive)	Crop protection relies mainly on human intervention with synthetic chemicals (more curative)

Important Sustainable agriculture practices are:

1. Crop rotations and increasing crop diversity.

2. Planting of cover crops and perennial crops..

3. Reduction or elimination of tillage operation.

4. Adoption of integrated pest management (IPM).

5. Integration of crops and livestock.

6. Adoption of different land use systems or agroforestry practices.

7. Scientific management of whole systems and landscapes.

Sustainable agriculture maintains a parity between the increasing pressure of food and / or feed demand and food and / or production in the future.

Sustainable farming minimises the use of toxic agricultural chemicals or pesticides that can harm the health of farmers and consumers.

i. Sustainable agriculture can prevents pollution.

ii. Sustainable agriculture can soil erosion.

iii. Sustainable agriculture can encourages biodiversity.

iv. Sustainable agriculture can saves energy.

Management of plant nutrients in sustainable agriculture

Adverse effects are being noticed due to the excessive and imbalanced use of synthetic or commercial fertilizers. Thus in sustainable or eco-friendly farming harmless bio-fertilizers are used along with organic manures. Use of bio-fertilizers in crop cultivation help in safeguarding the soil health and also the quality of crop products.

Bio-fertilizers are ready to use live materials containing beneficial microorganisms which on application to seed, root or soil mobilize the availability of plant nutrients by their biological activity or fix atmospheric nitrogen, and in turn improve the soil health or soil fertility.

Use of bio-fertilizers is increasing day by day and it has immense importance in sustainable agriculture. Use of various bio-fertilizers or microbial inoculants for management of major nutrients such as nitrogen and phosphorus are necessary for sustainability. The term '**Bio-fertilizer**' specifies input to meet the nutritional requirements of a crop through microbiological means. These bio-fertilizers are usually carrier based microbial preparations containing beneficial microorganisms in a viable state intended for seed / seedling or soil application, which enhance plant growth through nutrient uptake and / or growth hormone production. Important and popular microbial inoculants in our country are those that supplement nitrogen, phosphorus and plant growth promoting rhizobacteria (PGPR).

Beneficial effects of PGPR are as follows:–

i. Fixation of atmospheric nitrogen that is transferred to the plant.

ii. Solubilization of phosphorus.

iii. Production of plant hormones like IAA, GA3, cytokinin and induce formation of ethylene. Reduces deleterious effects of pathogens on crop growth by protection against pathogens by production of antibiotics.

iv. Solubilization of mineral nutrients by inducing specific ion flux in plant cell.

Mycorrhiza is a symbiotic association between plant roots and a few fungi. The fungal partner is benefited by obtaining its carbon requirements from host's photosynthates and the plant in turn gains the much needed nutrients especially phosphorus, calcium, copper and zinc which would otherwise be inaccessible to the host. This uptake of nutrients is facilitated with the help of a fine absorbing hyphae of the fungus. These fungi are associated with majority of agricultural crops.

Advantages of bio-fertilizers

i. Provide plant nutrients at very low cost.

ii. Made atmospheric nitrogen available to plants.

iii. Reduce the use of synthetic fertilizers.

iv. Solubilise phosphate and sulphate and increase uptake efficiency of plants.

v. Enhance plant growth by release of plant growth substances like, vitamins, auxins and hormones.

vi. Increase crop yields.

vii. Improve physical, chemical and biological fertility / properties of soil through their sustained activities in soil

viii. Have no harmful effect on soil and environment (Pollution free and eco-friendly).

ix. Required in small quantity.

x. Provide residual effects for subsequent crops.

xi. May hasten seed germination, flowering and maturity of crops.

xii. Helps in recycling /decomposition of crop residues / organic waste.

xiii. Provide protection against drought and some soil borne diseases.

Types of available Bio-fertilizers

The use of bio-fertilizers is quite important while practicing the concepts of integrated plant nutrient management and organic farming. Some of the commonly used bio-fertilizers are as follows:

For Nitrogen	
	Rhizobium for legume and pulse crops
	Azotobacter/Azospirillum for non-legume crops
	Acetobacter for sugarcane only
	BGA and *Azolla* for low land or transplanted rice
For Phosphorous	
	Phosphate solubilizing micro-organism (PSM) or Phosphatika or PSB for all crops to be applied with *Rhizobium, Azotobacter, Azospirillum,* BGA, *Azolla* and *Acetobacter*
	Phosphate Mobilizing Microbes, Mycorrhizae or Vesicular Arbuscular Mycorrhiza (VAM)
For enriched compost	
	Cellulolytic fungal culture
	Phosphotika and *Azotobacter* culture

Application methods / techniques of bio-fertilizers

1. Seed treatment

2. Seedling treatment

3. Soil application

Precautions to be taken during use of bio-fertilizers

i. Use recommended bio-fertilizer for particular crop.

ii. Do not expose treated seeds to direct sunlight.

iii. Sow treated seeds / Transplant treated seedling as early as possible (within 2-3 hours).

iv. Do not mix synthetic fertilizers, insecticides or pesticides directly with bio-fertilizers.

v. Use double dose of bio-fertilizers, if seeds are pre-treated with insecticides or pesticides or fungicides.

vi. Follow instruction given on packets.

vii. Check expiry date before use.

Use of bio-fertilizers is gaining momentum especially with emphasis on sustainable agriculture and organic farming. Use of bio-fertilizer and its benefits for sustainable agriculture depends on improving inoculant quality and effectiveness.

Sustainable crop protection measures:

In sustainable agriculture crop pest is managed by involving different techniques and the benefit is obtained as a cumulative effect of these techniques. In some respect pest management in sustainable agriculture is similar to IPM. The techniques generally taken are:-

1. Up to a certain limit (acceptable level or threshold limit) pest damage is allowed;

2. Predatory beneficial insects are encouraged to control pests;

3. Beneficial microorganisms or insects are encouraged to control pests (for this alternate host plants and /or an alternative habitat is provided - usually in a form of a shelterbelt, hedgerow, or beetle bank);

4. Crop and / or variety of crop is selected carefully (resistant varieties are chosen);

Basic objectives of pest management in sustainable crop production

1. To break pest cycles;

2. To promote diversity, and attract beneficial organisms;

3. To use biofumigants (mustard, broccoli, brassicas, or grasses like sudan grass, sorghum, sunn hemp, etc.).

4. Use of biofumigation is a strategy which uses plants, mainly belong to *Brassicaceae* family, that produce chemicals called glucosinolates, and when these glucosinolates are broken down by an enzyme to release various chemicals - including volatile isothiocynates (ITC's) which are toxic to soil borne pests and suppress numerous pests, including bacteria, fungi and nematodes and other disease pathogens. These biofumigants are less toxic and they persist less in the environment than synthetic fumigants.

In sustainable agriculture pest control involves the cumulative effect of many techniques, including:

1. Allowing for an acceptable level of pest damage;

2. Encouraging predatory beneficial insects to control pests;

3. Encouraging beneficial microorganisms and insects; this by serving them nursery plants and/or an alternative habitat, usually in a form of a shelterbelt, hedgerow, or beetle bank

4. Careful crop selection, choosing disease-resistant varieties

5. Growing / Planting companion crops that discourage or divert pests;

6. Rotating crops from season to season or year to year to interrupt pest reproduction cycles;

7. Using insect traps to monitor and control insect populations;

8. Using row covers to protect crops during pest migration periods;

9. Using pest regulating plants and biologic pesticides and herbicides;

10. Using reduced and no-till farming techniques (stale seed bed technique) as false seedbeds.

Each of these techniques also provides some other benefits like, soil protection and improvement, fertilization, pollination, water conservation, increase in cropping intensity or extension of crop growing period / season, etc.—and these benefits are both complementary and cumulative in overall effect on farm health. Effective organic pest control requires a thorough understanding of pest life cycles and interactions.

Crop protection methods used in sustainable agriculture are:

1. Identification and monitoring of crop pests;

2. Tactics used for pest prevention and suppression

i) Cultural pest control measures;

ii) Mechanical and physical pest control measures;

iii) Biological pest control measures; and

iv) Use of bio-pesticides.

Pest management in sustainable agriculture is similar to integrated pest management (IPM) in some respects.

Botanicals (also known as botanical pesticides or plant pesticides)

Botanicals in agriculture refer to all those types of products derived from plant sources, with potential for controlling weeds, phytophagus insects, plant diseases, etc. Basically it is a plant or plant part valued for its medicinal or therapeutics or pesticidal properties. Botanical pesticides are naturally occurring chemical derivatives of plants that act as antifeedants, discourage/ deter pests from approaching or settling (act as repellents), attractants, and/ or growth inhibitors. Plant parts generally used to make botanicals include stems, barks, leaves, roots, flowers, fruits, seeds, and rhizomes.

Use of botanicals is environmentally friendly pest management strategy. Ideally, botanicals are low-cost, non-toxic (or at least less toxic), non-persistent in the environment and preferably locally available. Examples of some common botanicals are tobacco, neem, pyrethrum, etc.

Properties of ideal 'botanicals'

1. Eco-Safety (safe for plants and animals).

2. Target specificity.

3. The active ingredient should be effective at low dose/rate.

4. Should be readily available and perennial in nature.

5. The plant parts to be used should be removable: leaves, flower and fruits.

6. The plants should require small space, less management, little water and fertilization.

7. Economic isolation procedures for the active components (less cost is to be incurred for the isolation).

8. Environment Friendly / Environment compatibility.

9. Lower exposure to pests.

10. Supplemental role to chemical pesticides enabling their use in integrated pest management and acceptability for use in sustainable agriculture.

Reasons for limited acceptance of botanicals

1. Lack of awareness.

2. Botanicals do not have instant knockdown effect.

3. Availability of competing products (newer synthetics, fermentation products, or microbial products)

4. Lack of better formulation.

Bio-control

Bio-control is defined as use of one organism to reduce or eliminate the population of another organism. The more appropriate term is 'Bio-control agents' and is defined as "biological derived agent or identical to a biological derived agent" and the term covers all types of environmentally safe products.

Bacteria, viruses, protozoa and fungi are the primary groups of microorganisms known to reduce insect populations; they often occur naturally in fields and function as components of biological control. The insect and mite control potential of natural and biological toxins is well known. Beneficial nematodes are also being used for pest control, especially against soil insects. The isolation of toxic metabolic compounds from microorganisms and their use is common and popular although their storing, successful marketing and distribution is not easy.

Bio-Pesticides

Bio- pesticides are derived from animals, plants, bacteria, and certain minerals, e.g., canola oil and baking soda, NPV, etc.

1. Derived from natural materials:

a) Microbial Pesticides – Bacteria (Bt), fungi, virus and nematodes

b) Bio-Chemical Pesticides – Comprised of naturally occurring substances that control pests by non-toxic mechanisms such as pheromones or some insect growth regulators. Bio- Pesticides must be registered with the Environmental Protection Agency (EPA).

2. Bio-pesticides, comprising living organisms or natural products derived from them are exemplified by:-

c) Plants – example: Pyrethrum (*Chrysanthemum* sp.), neem (*Azadirachta* or *Melia* sp. etc.)

d) Macrobials (example: *Trichogramma parasitoid*- a protozoan, *Cryptolaemus montrouzieri*- a coccinellid predator etc.), microscopic animals (example: Nematodes)

e) Microorganisms including bacteria (example: *Bacillus thuringiensis*), viruses (example: NPV or nucleopolyhedrosis virus)

f) Fungi (example: *Beauveria* sp.)

g) Transgenic plants containing a pest combating gene (example: Bt cotton, Bt brinjal, Bt tomato)

Bio-pesticides commonly used in India

1. *Bacillus thuringiensis,*

2. *Trichoderma viride,*

3. NPV (nucleopolyhedrosis virus) of *Helicoverpa armigera* and *Spodoptera litura,* and

4. Azadirachtin (Neem based pesticides).

Advantages of Bio-pesticides

a. Less harmful than conventional pesticides;

b. Generally affect only the target pest;

c. Effective in small quantities and often decompose quickly, leading to less pollution;

d. Bio-pesticides are excellent resistance-management tools;

e. Typically have no pre-harvest interval (PHI);

f. Can be approved for use in organic farming.

Bio-fungicide

Bio-fungicides are formulations of living organisms that are used to control the plant pathogenic fungi and bacteria. Bio-fungicide is selective and safe for humans and animals.

Some examples of bio-fungicides

Trichoderma harzianum, a species genetically engineered at Cornell, USA.

AQ10 has *Ampelomyces quisqualis* and is used to treat powdery mildew.

Fusaclean has *Fusarium oxysporum* and is used to treat wilt caused by other species of *Fusarium*.

Rotstop has *Peniophora gigantea* and is used to control butt rot (a fungal disease of plants) caused by *Heterobasidion annosum*.

Several bio-pesticides, e.g., *Trichoderma viridi, Bacillus thuringiensis* (Bt), Nuclear Polyhedrosis Viruses (NPV), GV etc., botanical pesticides (Neem), bio-control agents *(Trichogramma, Cryptolaemus, Chrysoperla* etc.) are capable of controlling pests and diseases effectively.

Sustainable Methods of Weed Management

Weeds are problematic and major pests of field crops.

» Weeds can reduce crop yield.

» Weeds can increase the cost of crop cultivation and can also increase the processing cost.

» Weeds can reduce the quality of the produce.

In sustainable agriculture weeds have a huge impact on crops where use of toxic chemicals (herbicides) is restricted (or prohibited).

Tips for effective weed management in sustainable agriculture

1. Sanitation and clean cultivation.

2. Selection of appropriate variety

3. Sowing of clean and good quality seeds

4. Cultivation of good shading crops or cover crops.

5. Effective and judicious use of water and fertilizer is the key to manage weeds. Fertilizer (particularly nitrogen fertilizer) and irrigation water should be used carefully. Weed growth can be checked by using drip irrigation.

1. Thermal weed control method

In this method flaming equipment (weed flamer) is used which creates direct contact between the flame and the weed plants, as a result there is rupturing the flame and the weed plants, as a result there is rupturing of plant cells. Sometimes heat from steam and boiling water are used.

Sometimes thermal control involves the outright burning down of the weeds.

Flaming can be used either before crop emergence to give the crop a competitive advantage   or after the crop has emerged (however, flaming at this point should done carefully otherwise it may damage the crop).

Although the initial equipment cost for weed flamer may be high, flaming for weed control may prove cheaper than hand or manual weeding.

Weed Zapper

Electricity is a thermal weed management method which is used in weed zapper. The electricity is applied directly to the plant and once it gets inside, travels to the roots and comes out through the soil, and effectively kills the weeds. Electric zapper uses thermogenic (heat) energy to stop normal plant functions and kill the weeds. This control measure is eco-friendly and does not disturb the soil.

2. Soil solarization

During hot summer months, farmers sometimes sterilize their soil through solarization [a transparent/clear plastic film/sheet is placed over an area (field) after it has been tilled, and the heat created under the plastic film/ sheet (which is tightly sealed at the edges), becomes intense enough to kill weed seeds].

3. Mulch

A mulch is a layer of material applied to the surface of soil. Mulching or covering the soil surface can prevent germination of weed seed by blocking light transmission (this prevents germination of weed seeds).

Allelopathic chemicals in the mulch (e.g., β-pinene and camphene) also can suppress emergence of weed seedlings.

There are many forms of mulches available, however, three common ones are:-

i. Live mulch / Living mulch

A living mulch is usually a plant species that grows densely, quickly and low to the ground, such as rice bean, cowpea. Living mulches can be planted before or after a crop is established. It is important to kill, till in, or otherwise manage the live mulch so that it does not compete the main (actual) crop.

ii. Organic mulches

Materials like straw, leaves, bark, and composted material are used as organic mulch and can provide effective weed control. Producing the material on the farm depending on the amount needed to suppress weed emergence or weed growth. Use of on farm product is recommended since the cost of purchased materials (mulches) may increase the cost of production and is not sustainable.

An effective but labour-intensive system is the uses of newspaper and straw (two layers of newspaper are placed on the ground, followed by a layer of straw / hay). It is important to make sure that the mulching material (straw or hay) does not contain any weed seeds.

iii. Inorganic mulch

Materials such as polythene sheet (preferably black polyethylene) are been used for weed management in a wide range of crops.

4. Mechanical weed management

It is one of the most effective methods for managing weeds (although it is both time consuming and labour- intensive). The choice of implement, timing, and frequency depend on the crop growth stage (crop canopy or structure and form of the crop) and the type and number of weeds.

Tillage or soil cultivation involves killing emerging weeds or burying freshly shed weed seeds below the depth from which it will difficult for them to germinate.

The soil seed-bank is the reserve of weed seeds present in the soil. It is important to remember that any ecological approach to weed management begins and ends in the destruction of soil seed- bank (of weeds).

Farmers should take a close observation about the composition of the seed-bank and this can help them to select the best option and make practical weed management decisions in a sustainable manner..

5. Stale seedbed technique

The stale or false seedbed technique of flushing out weed seeds from the soil works by depleting the seed-bank.

In this technique soil is tilled / cultivated 15-20 days before sowing of crop seeds. Within this period (after cultivation of land and 2-3 weeks before sowing of crop seeds), weeds are emerged and emerging weeds are killed by any suitable method or by tillage operation.

This technique helps to reduce / destruct the weed seed-bank and reduces subsequent emergences of weeds.

6. Crop rotation

The goal of a crop rotation is to create an unstable or unfavourable environment that discourages weeds from becoming established in the field. A farmer must take into account - the type of soil, soil fertility, the weather / climate, nutrient management practices and the sequence of crops for successful management of weeds.

7. Allelopathy

Allelopathy is a biological phenomenon by which a plant (or an organism) influence the germination, growth, survival, and reproduction of neighbouring plants / other plants (organisms) by producing **allelochemicals** or **bio-chemical(s)**. These chemicals can have beneficial (positive allelopathy) or detrimental (negative allelopathy) effects on the target plants / organisms and the community.

Allelopathy is an important component of biological control.

Allelopathy is an alternative and organic approach to weed control that uses chemicals that are excreted from a plant to cause either direct or indirect harm to weeds by negatively affecting their germination, growth, or development.

Nearby weeds can be affected by allelopathic chemicals entering the rhizosphere from the roots or the aerial parts of the crop plant.

Crop residues from cover crops, or other organic mulches can also be used to suppress weeds through such allelopathic interactions.

8. Biological weed management

Biological weed management involves using living organisms, such as insects (bugs), bacteria, virus, nematodes or fungi (even the bio-control agents may be grazing animals), to reduce weed populations. Biological control agents provide a more eco-friendly, self-sustaining and cost-effective alternative to chemical control. However they suppress, rather than eliminate, a weed population. In biological management total / complete eradication of weeds is not possible.

Biological weed control involves the release of organisms that attack plants to control weeds. The aim of biological control is to shift the balance of competition between the weed and the crop in favour of the crop and against the weed.

The biological control agent, normally a fungus or insect, may not necessarily kill the target weed but should, at the least, reduce its vigour and competitive ability.

Predatory or parasitic micro-organisms, insects or other animals like fishes, mammals, and snails are used to manage weed populations.

Plants as intercrop may be used (e.g., cowpea as intercrop may effectively reduce the growth of weeds in sorghum).

From a practical point of view the organism or agent should prevent the weed setting seed or producing other reproductive parts.

9. Chemical weed management

Some less toxic or plant origin herbicides or **organic herbicides** (such as vinegar, lemon juice, corn gluten, citric acid, clove oil and acetic acid) are allowed and used in sustainable and organic agriculture. It is recommended that farmers should try the other techniques (mentioned above) first and should use organic herbicides as a last resort (used only when all else has failed).

Organic herbicides kill only contacted tissue so good spray coverage is essential. Organic herbicides work when enough volume and concentration is used. However, these herbicides are expensive and may not be affordable for commercial crop production.

Herbicidal soaps (that contain fatty acids), industrial vinegar (which contains much higher levels of acetic acid). Acid-based herbicides burn down some young weeds. Corn gluten meal can kill grass weeds and broadleaf weeds. These organic products are effective in controlling weeds when the weeds are small but are less effective on older plants.

Examples of some organic herbicides (available in foreign markets):

» Matratec (50% clove oil),

» Weed Pharm (20% acetic acid),

» WeedZap (45% clove oil + 45% cinnamon oil),

» C-Cide (5% citric acid),

» GreenMatch (55% d-limonene), and

» GreenMatch EX (50% lemongrass oil).

Chapter - 17

Impact of Plant Health Measures on Human Health

Both positive and negative impact has been noticed on the human health for the measures taken to keep plants healthy. In spite of knowing the importance of plant health practices on human health; we often exclude it from coordination amongst various sectors under 'One Health' banner. Long term usage of chemical pesticides used for the control of plant pests can affect the human health with an increment in the cancer risk along with disruption of immune as well as hormone system function & impaired development of brain (Gilden *et al.*, 2010). On other hand, abrupt insecticides usage makes it more difficult to control due to development of resistance among vectors of human diseases such as mosquitoes (Yadouleton *et al.*, 2009). There may be occurrence of antimicrobial-resistant infections in humans due to antimicrobials used in crop production along with antimicrobial resistance genes and pathogens present in animal manure transmitted via the food chain (Jiang *et al.*, 2015; Checcucci *et al.*, 2020). We have noticed a common alignment of ecosystem health, human health and farmer's incentives for many plant health practices. Though there is a requirement of additional labour, which may inhibit the process of adoption but composting and anaerobic digestion of manure shows a reduction of pathogen prevalence and antimicrobial resistance risk with the increment in the effectiveness of manure as a fertilizer (Millner *et al.*, 2014; Ndambi *et al.*, 2019). The use of wood and fossil fuels for cooking gets highly reduced with biogas produced through anaerobic digestion along with saving of time and money of the farm families and also reduction of smoke inhalation by women. Dangerous mycotoxins produced from fungal species can lead to contamination of crops which can be prevented with a strong plant health (Strosnider *et al.*, 2006). Motivated by the biological vulnerability of children and women's primary responsibility for child health in many societies, ongoing research investigates the role of gender in mediating adoption of mycotoxin control measures (Bauchet *et al.*, 2021). Absorption of heavy metals, waste disposal, and pesticides, present in irrigation water due to upstream industrial activities or soils along with the veterinary medicine used in livestock production occurs in food crops, contributing to intellectual disability & ill-health (Gibb *et al.* 2019). Heavy metals uptake in plants can be greatly reduced with the addition of organic matter having a beneficial impact on plant health (Sharma and Nagpal, 2018; Park *et al.*, 2011).

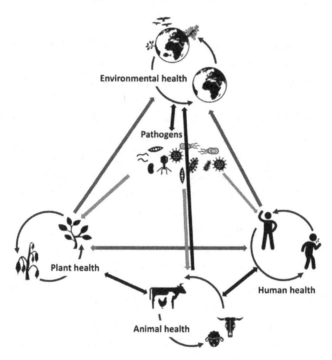

There has been clear indication of trade-offs between public good & farmers' incomes due to practices, such as injudicious antimicrobial use or low-cost but highly toxic pesticides. Policy options in such cases include rewarding value chain actors for responsible practices, restriction in the access to inputs which may expose One Health at risks, or designing and enforcing appropriate food safety and environmental regulations. Avoidance of punitive enforcement rules along with reflection of the global shift in regulatory practice toward proactive management of risk (Blanc, 2018) must be feasible for implementation by the farmers. Ensuring the quality and safety of available agrochemicals & capacity building along the value chain must be prime role of state.

Achieving food security while minimizing environmental harms

'One Health' frames the complex interaction between plant health and human wellbeing at a higher level through its environmental dimension. Food security can be achieved by increasing crop yields through healthy plants for feeding the growing global population. However, environmental processes that impact the human health is hampered by various agricultural production. 34% of greenhouse gas emissions in the atmosphere, 84% of fresh water consumption (Shiklomanov and Rodda, 2003) along with biggest source of N and P pollution in aquatic systems (Galloway *et al.* 2008; Carpenter and Bennett, 2011) is contributed by agriculture (Crippa *et al.*, 2021). Further, events like soil erosion and degradation (Jie *et al.*, 2002; Montgomery, 2007) & encroachment on natural areas resulting in biodiversity loss (Norris, 2008), thereby increasing the risk of new emerging zoonotic pathogens (Gibb *et al.* 2020) is led due to agriculture activities. Thus, One Health necessitates for a sustainable agricultural intensification.

There are two approaches for achieving sustainable agricultural intensification:

1. Rationalized use of pesticides and fertilizers along with proper soil and water conservation measures helps in improving the efficiency of inputs;

2. Breaking from trends of increasing industrialization, specialization and commercialization and moving towards integration of livestock and crops (Ramankutty *et al.*, 2018) or by redesigning the system like usage of agro-ecological or organic principles (Pretty *et al.*, 2018).

If there is an increment in the efficiency of conventional systems by the conversion of additional land into agriculture, food production may increase by 30% per land area (Mueller *et al.* 2012), but will result in higher water usage, greenhouse gas emissions and pollution and negative impacts on biodiversity. On the contrary, organic systems will result in less environmental damage within cultivated areas (Seufert and Ramankutty 2017) but a reverse trade-off might occur with 19–25% lower food production per land area. Across ethnic groups, localities and along the agricultural value chains, there has been a great diversity of gender roles and constraints which need to be addressed for empowering female stakeholders (FAO 2010; Olawoye 2018). Conduction of training at times when needed and can be attended by women with increment in the ranks of female extension workers show a positive response for active participation by women (Quisumbing and Pandolfelli, 2010).

Chapter - 18

Conclusion

'One Health' is a multi-disciplinary and collaborating approach that works at the global, national, regional and local levels with the goal to achieve better health for all, that means human being, animals, plants and their surroundings or environments. Worldwide, 820 million people are food insecure, with chronic hunger and malnutrition. To ensure food security, eliminate hunger, and alleviate poverty, sustainable agriculture must be the foundation. Over the previous decade, tremendous progress has been made in reducing undernourishment. The Food and Agriculture Organization of the United Nations (FAO), on the other hand, sees extreme weather and climatic variability as a "key cause" of recent rises in world famine. Tornadoes, floods, droughts, wildfires, blizzards, new illnesses, and earthquakes all have a significant influence on food security by affecting agriculture production, availability of food, access to and consumption of food, and food stability. Any of these factors can contribute to disease outbreaks in people and animals (IFAD, UNICEF, WFP, WHO and FAO, 2018). Poverty-stricken people in both rich and developing nations are especially vulnerable to fluctuations in food costs, natural catastrophes, and trans-boundary animal illness, which are the most serious risks to food security in the twenty-first century (FAO, 2015). As a result, One Health is a critical way to mitigating these issues. Agriculture has a critical role in preserving the various ecological balances associated with it. And these roles cannot be fulfilled without the assistance of an agronomist. From caring for degraded soil to advising and teaching farmers, from introducing excellent agricultural techniques to new technology, from recognizing the strengths, weaknesses, fit's, etc. of various practices and goods to making decisions towards that integrated approach, agronomists are inextricably linked to all aspects of agriculture as well as the Earth itself. They are the grassroots workers who are making a significant contribution to the 'One Health' strategy. Some attempts have been suggested in order to identify the shortcomings and take some measures ahead. It's high time that we need to join together and act properly to make this movement a successful one and soon we will be able to attain our objective of "One World, One Health ". The ultimate goal is to promote cross-disciplinary research collaborations and knowledge exchange at several levels in order to enhance ecosystem health, safeguard and uphold the wellbeing of all animals. **World 'One Health' Day** is observed on **November 3** in order to promote awareness across the globe. **Annual 'One Health' Day** is observed to educate people and increase awareness among partners working in human, animal and environmental health because working together help us to have the highest impact on improving health for people, animals, plants and our surrounding environment. Urgent attention and prompt action is needed to keep the planet and humans healthy.

Dismantling the disciplinary boundaries that still exist between ecological, evolutionary, and environmental sciences and human and veterinary health is urgently needed for the viability or sustainability of the 'One Health' idea.

"The first wealth is health," so care must be taken for effective and adequate measures for our health and we should always eat healthy. For achieving the optimal health we should recognize the interconnection between people, animals, plants, and their shared environment (ecology).

References

Abbott, I., (2006). Mammalian faunal collapse in Western Australia, 1875–1925: the hypothesised role of epizootic disease and a conceptual model of its origin, introduction, transmission and spread. *Australian Journal of Zoology*, **33**: 530–561.

Ahmed, S. (2019). National Institute of One Health to Come up on MAFSU Campus. Available from: https://www.timesofindia.indiatimes.com/city/nagpur/natl-institute-of-one-health-to-come-up-on-mafsu-campus/articleshow/70899851.cms. Retrieved on 30-05-2020.

Anonymous (2021). DBT Launched First 'One Health' Project. Web: https://www.nextias.com/current-affairs/14-10-2021/dbt-launched-first-one-health-project [Accessed on 31st January, 2023].

Anonymous (2019). Bionity (2019). Multidrug-resistant bacteria: urban brown rats as possible source. Web: https://www.bionity.com/en/news/1162662/multidrug-resistant-bacteria-urban-brown-rats-as-possible-source.html (Veterinärmedizinische Universität Wien) [Accessed on 31st January, 2023].

Anonymous (2022a). One Health Commission (2022). What is One Health? Web: https://www.onehealthcommission.org/en/why_one_health/what_is_one_health/ [Accessed on 31st January, 2023].

Anonymous (2022b). Centers for Disease Control and Prevention (2022). *One Health Basics*. Web: https://www.cdc.gov/onehealth/basics/index.html [Accessed on 31st January, 2023].

Anonymous (2022c). Crack it Today Affairs (2022). One Health Framework: Uttarakhand. Web: https://crackittoday.com/current-affairs/one-health-framework-uttarakhand/ [Accessed on 31st January, 2023].

Anonymous (2022d). Drishti IAS. (2022). One Health Concept. Web: https://www.drishtiias.com/daily-updates/daily-news-analysis/one-health-concept-1/print_manually [Accessed on 31st January, 2023].

Anonymous (2019). Public Health Notes (2019). 'One Health'- A Multi-Dimensional Approach to Health. Web: https://www.publichealthnotes. com/1332-2/ [Accessed on 31st January, 2023].

Asaaga, F. A., Young, J. C., Oommen, M. A., Chandarana, R., August, J., Joshi, J., Chanda M.M., Vanak A.T., Srinivas P.N., Hoti S.L., Seshadri T., and Purse, B. V. (2021). Operationalising the "One Health" approach in India: facilitators of and barriers to effective cross-sector convergence for zoonoses prevention and control. *BMC Public Health*, **21** (Article No. 1517): 1-21.

Bauchet J, et al. (2021). Impacts of pre-harvest and post-harvest treatments on reducing aflatoxin contamination in smallholder farmers' maize. AEA RCT Registry. 2021. https://doi.org/10.1257/rct.7067-1.0 [Accessed on 31st January, 2023].

Benson, Todd (2006). Agriculture and Health in the Policymaking Process. **In**: Understanding the links between agriculture and health. International Food Policy Research Institute, Washington D.C., USA.

Bernard, A., Broeckaert, F., De Poorter, G., De Cock, A., Hermans, C., Saegerman, C., and Houins, G. (2002). The Belgian PCB/dioxin incident: Analysis of the food chain contamination and health risk evaluation. *Environmental Research*, **88**(1), 1-18.

Blanc F. (2018). From chasing violations to managing risks: origins, challenges and evolutions in regulatory inspections. Edward Elgar Publishing. https://doi.org/10.4337/9781788112499.

Boliko, M. C., (2019). FAO and the situation of food security and nutrition in the world. *Journal of nutritional science and vitaminology*, **65** (Supplement): S4-S8.

Bos, Robert (2006). Opportunities for Improving the Synergies between Agriculture and Health. **In**: Understanding the Links between Agriculture and Health. International Food Policy Research Institute, Washington D.C., USA.

Brahmaprakash, G.P., Girisha, H.C., Navi, V. and Hegde, S.V. (2004). Biological nitrogen fixation in pulse crops. In: *Pulses in New Perspectives* (eds.) M. Ali, B.B. Singh, Shiv Kumar and Vishwa Dhar). Indian Society of Pulses Research and Development, Kanpur, India, pp. 271-286.

Calistri, P., Iannetti, S., L. Danzetta, M., Narcisi, V., Cito, F., Di Sabatino, D., Bruno R., Sauro F., Atzeni M., Carvelli A., & Giovannini, A. (2013). The components of 'one world–one health'approach. *Transboundary and emerging* disease,s **60**: 4-13.

Cameron, S.A., Lozier, J.D., Strange, J.P., Koch, J.B., Cordes, N., Solter, L.F., & Griswold, T.L. 2011. Patterns of widespread decline in North American bumblebees. *Proceedings of the National Academy of Sciences*, **108** (2): 662–667.

Carpenter, S.R. and Bennett, E.M. (2011). Reconsideration of the planetary boundary for phosphorus. *Environmental Research Letters*, 6(1): 014009.

Chatterjee, P., Kakkar, M. and Chaturvedi, S. (2016) Integrating one health in national health policies of developing countries: India's lost opportunities. *Infectious Diseases of Poverty*, **5** (1): 87.

Checcucci A., Trevisi P., Luise D., Modesto M., Blasioli S., Braschi I., & Mattarelli P. (2020). Exploring the animal waste resistome: the spread of antimicrobial resistance genes through the use of livestock manure. *Frontiers in Microbiology*, **11**:1416.

Chien, Y. J. (2013). How did international agencies perceive the avian influenza problem? The adoption and manufacture of the 'One World, One Health'framework. *Pandemics and Emerging Infectious Diseases: The Sociological Agenda*, pp. 46-58.

Chirwa, P.W., & Adeyemi, O. (2020). Deforestation in Africa: Implications on food and nutritional security. *Zero hunger*, 197-211.

Crack it Today Affairs (2022). One Health Framework: Uttarakhand. Web: https://crackittoday.com/current-affairs/one-health-framework-uttarakhand/ [Accessed on 31st January, 2023].

Crippa, M., Solazzo, E., Guizzardi, D., Monforti-Ferrario, F., Tubiello, F.N.,

Leip, A. (2021). Food systems are responsible for a third of global anthropogenic GHG emissions. *Nature Food,* **2**(3):198–209.

Crisci, E. (2022). From Open Access to Circular Health: Ilaria Capua's Journey through Science and Politics. *Viruses,* **14** (6) 1296: 1-8.

Dasgupta, Rajib., Tomley, Fiona., Alders, Robyn., Barbuddhe, Sukhadeo B., and Kotwani, Anita. (2021). Adopting an intersectoral One Health approach in India: Time for One Health Committees. *Indian Journal of Medical Research,* **153** (3): 281-286.

DBT. (2020). Department of Biotechnology, National Expert Group on One Health. Available from:http://www.dbtindia.gov.in/sites/default/files/national%20expert%20group%20on%20one%20health.pdf. Retrieved on 18-06-2020.

DBT-One Health India Conference Declaration. (2019). Available from: https://www.img1.wsimg.com/blobby/go/bdf18a05-5d66-48cb-8cad-38ee88a6d879/downloads/ohic%202019_declaration.pdf?ver=1556639010884. Retrieved on 18-06-2020.

Deccan Herald (2021). World 'One Health' Day: Covid-19 helped cause but uncharted turf ahead. Web: https://www.deccanherald.com/opinion/world-one-health-day-covid-19 helped-cause-but-uncharted-turf-ahead-1046985.html [Accessed on 3rd March, 2023]

Delhi Declaration: 2019. (2019). Available from: https://www.who.int/docs/defaultsource/searo/whe/delhi-declaration-emergency-preparednesssouth-east-asia-region.pdf?sfvrsn=3293354d_2. Retrieved on 15-06-2020.

Destoumieux-Garzón, D., Mavingui, P., Boetsch, G., Boissier, J., Darriet, F., Duboz, P., Fritsch, C., Giraudoux, P., Roux, F. Le, Morand, S., Paillard, C., Pontier, D., Sueur, . and Voituron. Y. (2018). The One Health Concept: 10 Years Old and a Long Road Ahead. *Frontiers in Veterinary Science,* **5** (14): 1-13.

European Centre for Disease Prevention and Control, European Food Safety Authority and European Medicines Agency, 2015. ECDC/EFSA/EMA first joint report on the integrated analysis of the consumption of antimicrobial agents and occurrence of antimicrobial resistance in bacteria from humans and food-producing animals. *EFSA Journal*, **13** (1): 4006.

European Commission (2000). White paper on food safety. Commission of the European Communities, Brussels, 12 January, pp. 52 Web: http://ec.europa.eu/dgs/health_consumer/library/pub/pub06_en.pdf [Accessed on 5th August, 2012].

European Commission (2009). The Rapid Alert System for Food and Feed (RASFF). Annual Report, 2008. pp. 56.

Ewen, J.G., Acevedo-Whitehouse, K., Alley, M.R., Carraro, C., Sainsbury, A.W., Swinnerton, K., and Woodroffe, R. (2012). Empirical consideration of parasites and health in reintroduction. In: Ewen, J.G., Armstrong, D.P., Parker, K.A., Seddon, P.J. (Eds.), Reintroduction Biology. Wiley-Blackwell, Hoboken, NJ, USA, pp. 320–335.

Farm Management (2023). Popular Unsustainable Techniques Used in Modern Agriculture Web: https://www.farmmanagement.pro/popular-unsustainable-techniques-used-in-modern-agriculture/ [Accessed on 9th March, 2023].

Food and Agriculture Organization of the United Nations (2021). One Health. Web: https://www.fao.org/one-health/en/ [Accessed on 31st January, 2023].

Garcia, S., Osburn, B., and Jay-Russell, M. (2020). One Health for Food Safety, Food Security, and Sustainable Food Production. *Frontiers in Sustainable Food Systems*, **4**(1): 1-9.

Garg, M., Sharma, N., Sharma, S., Kapoor, P., Kumar, A., Chunduri, V., & Arora P. (2018). Biofortified Crops Generated by Breeding, Agronomy, and Transgenic Approaches Are Improving Lives of Millions of People around the World. *Frontiers in Nutrition*, **5**: 1-33.

Gibb, H.J., Barchowsky, A., Bellinger, D., Bolger, P.M., Carrington, C., Havelaar, A.H., Oberoi, S., Zang, Y., O'Leary, K. and Devleesschauwer, B. (2019). Estimates of the 2015 global and regional disease burden from four foodborne metals—arsenic, cadmium, lead and methylmercury. *Environmental Research*, **174**:188–194.

Gibb, R., Redding, D.W., Chin, K.Q., Donnelly, C.A., Blackburn, T.M., Newbold, T., Jones, K.E. (2020). Zoonotic host diversity increases in human-dominated ecosystems. *Nature*, **584**: 398–402.

Gilden, R.C., Huffling, K., and Sattler, B. (2010). Pesticides and health risks. *Journal of Obstetric, Gynecologic & Neonatal Nursing*, **39**(1):103–110.

Gills, M.S., Prasad, K. and Ahalawat, I.P.S. (2009). Improving sustainability of rice-wheat cropping system through pulses: weeds and imperatives. **In**: *Legumes for Ecological Sustainability* (eds.) M. Ali, S Gupta, P.S. Basu and Naimuddin. Indian Society of Pulses Research and Development, Kanpur. Pp. 71-91.

Harrington, L.A., Moehrenschlager, A., Gelling, M., Atkinson, R.P.D., Hughes, J., Macdonald, and D.W. (2013). Conflicting and complementary ethics of animal welfare considerations in reintroductions. *Conservation Biology*, **27**: 486 500.

Hawkes, C., and T. Ruel, M. (2006). Agriculture, Environment, and Health: Toward Sustainable Solutions. **In**: R. NUGENT, & A. DRESCHER, UNDERSTANDING THE LINKS BETWEEN AGRICULTURE AND HEALTH. Washington D.C.: International Food Policy Research Institute.

Hoffmann, V., Paul, B., Falade, T., Moodley, A., Ramankutty, N., Olawoye, J., Djouaka , R., Lekei, E., de Haan, N., Ballantyne, P. and Waage, J. (2022). A one health approach to plant health. *CABI Agriculture and Bioscience*, **3**(Article No. 62): 1-7.

IASGYAN (2021). ONE HEALTH CONCEPT. Web: https://www. iasgyan.in/blogs/one-health-concept [Accessed on 31st January, 2023].

ICRISAT. (2016). *Catch the Pulse*. Patancheru 502 324, Telangana, India: International Crops Research Institute for the Semi-Arid Tropics. pp. 36.

Indian Express (2021). Why India needs a 'One Health' vision to tackle the crisis caused by the pandemic. Web: https://indianexpress.com/article/ opinion/why-india-needs-a-one-health-vision-7261541/ [Accessed on 3rd March, 2023].

International Organization for Standardization (ISO) (1995). *ISO 8402 Quality Management and Quality Assurance. Vocabulary.* ISO, Geneva, pp. 28.

Jiang, X., Chen, Z. and Dharmasena, M. (2015). The role of animal manure in the contamination of fresh food. **In**: Advances in microbial food safety (Vol. 2). Woodhead Publishing Series in Food Science, Technology and Nutrition, pp. 312-350. https://doi.org/10.1533/9781782421153.3.312.

Jie, C., Jing-Zhang, C., Man-Zhi, T., Zi-tong, G. (2002). Soil degradation: A global problem endangering sustainable development. *Journal of Geographical Sciences*, **12**(2): 243–252.

Jovanovic, Nina, Ricker-Gilbert, Jacob, Ketiem, Patrick, Bauchet, Jonathan and Hoffmann, Vivia (2022). Impacts of pre-harvest and post-harvest treatments on reducing aflatoxin contamination in smallholder farmers' maize. AEA RCT Registry. 2021. https://doi.org/10.1257/rct.7067-1.0.

Kaul, R. (2020). Public Expenditure on Health to be In-Creased. Available from: https://www.hindustantimes.com/india-news/public-expenditure-on-health-to-be-increased-nirmala-sitharaman/story-dipsy19htlyycfvj03lkrj.html. Retrieved on 17-05-2020.

Mackenzie, J., and Jeggo, M. (2019). The One Health Approach—Why Is It So Important? *Tropical Medicine and Infectious Disease*, 4(2): 1-4.

Mackenzie, J.S., McKinnon, M. and Jeggo, M. (2014). One health: From concept to practice. **In**: Confronting Emerging Zoonoses: The One Health Paradigm. Springer Japan, 2014 Available from: https://www.link.springer.com/chapter/10.1007 / 978-4-431-55120-1_8. Retrieved on 15-06-2020.

MacPhee, R.D.E., and Greenwood, A.D. (2013). Infectious disease, endangerment, and extinction. *International Journal of Evolutionary Biology*, 2013 (**571939**): 1-9.

Mantovani, A. (2008). Human and veterinary medicine: the priority for public health synergies. *Veterinaria. Italiana*, **44**: 577-582.

Mathews, F. (2009). Zoonoses in wildlife: integrating ecology into management. *Advances in Parasitology*, **68**: 185–208.

Mfutso-Bengu, J.M., and Taylor, T.E. (2002). Ethical jurisdictions in biomedical research. *Trends in Parasitology*, **18**(5): 231-234.

Millner, P., Ingram, D., Mulbry, W. and Arikan, O.A. (2014). Pathogen reduction in minimally managed composting of bovine manure. *Waste Management*, **34**(11):1992–1999.

Mint (2021). Centre plans labs in every block. Web: https://www.livemint.com/politics/policy/centre-plans-labsin-every-block-11589745155999.html [Accessed on 12th March, 2021].

Montgomery, D.R. (2007). Soil erosion and agricultural sustainability. *Proceedings of the National Academy of Sciences*, **104**(33): 13268–13272.

Mudryj, A.N., Yu, N. and Aukema, H.M. (2014). Nutritional and health benefits of pulses. *Applied Physiology, Nutrition, and Metabolism*, **39**(11):1197-1204.

Mueller, D.M., Gerber, J.S., Johnston, M., Ray, D.K., Ramankutty, N., Foley, J.A. (2012). Closing yield gaps through nutrient and water management. *Nature*, 490: 254–257.

National Centre for Disease Control (2020). Directorate General of Health Services, Ministry of Health and Family Welfare, Government of India. Available from: https://www.ncdc.gov.in/ index1.php? lang =1&level =1&sublinkid=144&lid=152. Retrieved on 16-06-2020.

Ndambi, O.A., Pelster, D.E., Owino, J.O., de Buisonje, F. and Vellinga, T. (2019). Manure management practices and policies in Sub-Saharan Africa: Implications on manure quality as a fertilizer. *Frontiers in Sustainable Food Systems*. Volume 3 – May 2019 (Article 29).

Norris K. (2008). Agriculture and biodiversity conservation: opportunity knocks. *Conservation Letters*, **1**(1):2–11.

Park, J.H., Lamb, D., Paneerselvam, P., Choppala, G., Bolan, N., Chung, J.W. (2011). Role of organic amendments on enhanced bioremediation of heavy metal (loid) contaminated soils. *Journal of Hazardous Materials*, 185(2–3):549–554.

Paul, B.K., Butterbach-Bahl, K., Notenbaert, A., Nderi, A.N., Ericksen, P. (2021). Sustainable livestock development in low and middle income countries—shedding light on evidence-based solutions. *Environmental Research Letters*, **16** (1):011001.

Pinstrup-Andersen, P. (2010). The African Food System and Its Interaction with Human Health and Nutrition. New York: Cornell University Press.

Prejit, N. (2018). Development of a novel course and new centre dedicated to "one health" to support the control of zoonosis and other public health needs of the country. *Journal of Foodborne Zoonotic Disease*, **6**(2): 18-22.

Prejit, N. and Jess, V. (2018). The "COHEART" Journey. Kerala Veterinary and Animal Sciences University, Kerala. Available from: http://www. coheart.ac.in/ uploads/ about/1554357111.pdf. Retrieved on 18-06-2020.

Pretty, J., Benton, T.G., Bharucha, Z.P., Dicks, L.V., Dicks, L.V., Flora, C.B., Godfray, H.C.J., Goulson, D., Hartley, S., Lampkin, N., Morris, C., Pierzynski, G., Prasad, P.V.V., Reganold, J., Rockstrom, J., Smith, P., Thorne, P., Wratten, S. (2018). Global assessment of agricultural system redesign for sustainable intensification. *Nature Sustainability*, **1**:441–446.

Queenan, K., Häsler, B., and Rushton, J. (2016). A One Health approach to antimicrobial resistance surveillance: is there a business case for it? *International Journal of Antimicrobial Agents*, **48** (4): 422-427.

Quisumbing, A.R. and Pandolfelli, L. (2010). Promising approaches to address the needs of poor female farmers: resources, constraints, and interventions. *World* Developmen,t **38**(4):581–592.

Ramankutty, N., Mehrabi, Z., Waha, K., Jarvis, L., Kremen, C., Herrero, M., Rieseberg, L.H. (2018). Trends in global agricultural land use: implications for environmental health and food security. *Annual Review of Plant Biology*, **69**(1):789–815.

Ranjalkar, J. and Sujith, J.C. (2019). India's national action plan for antimicrobial resistance-an overview of the context, status, and way ahead. *Journal of Family Medicine and Primary Care*, **8**(6): 1828-1834.

100

Robinson, R.A., Lawson, B., Toms, M.P., Peck, K.M., Kirkwood, J.K., Chantrey, J., Clatworthy, I.R., Evans, A.D., Hughes, L.A., Hutchinson, O.C., John, S.K., Pennycott, T.W., Perkins, M.W., Rowley, P.S., Simpson, V.R., Tyler, K.M., and Cunningham, A.A. (2010). Emerging infectious disease leads to rapid population declines of common British birds. PLOS One, 5(8): 1-12.

Rose, T. J., Hardiputra, B., and Rengel, Z. (2010). Wheat, canola and grain legume access to soil phosphorus fractions differs in soils with contrasting phosphorus dynamics. *Plant Soil*, **326**: 159–170.

Samanta, I. and Bandyopadhyay, S. (2019). Antimicrobial resistance: one health approach. **In:** I. Samanta, and S. Bandyopadhyay, Antimicrobial Resistance in Agriculture: **perspective, policy and mitigation**, pp. 365-372, Academic Press (ELSEVIER).

Sasson, A. (2012). Food security for Africa: an urgent global challenge. *Agriculture and Food Security*, **1**(1): 1-16.

Sengupta, Kajal (2022). Climate Smart Agriculture and Pulse Cultivation. **In:** Book entitled 'Climate Change Dimensions and Mitigation Strategies for Agricultural Sustainability', Vol. II, Chapter 2 (Editors: Suborna Roy Choudhury & Chandan Kumar Panda, *New Delhi Publishers*, New Delhi, pp. 11-18.

Sengupta, Kajal and Biswas, S. (2017). Pulse Production and Ecology: The Issues of Community Mobilisation in India. *Agricultural Extension Journal*, **1**(1): 31- 34.

Seufert, V., Ramankutty, N. (2017). Many shades of gray—The context-dependent performance of organic agriculture. *Science Advances*, **3**(3) (e1602638).

Sharma, A., Nagpal, A.K. (2018). Soil amendments: a tool to reduce heavy metal uptake in crops for production of safe food. *Reviews in Environmental Science and Bio/Technology*, **17**(1):187–203.

Shiklomanov, I.A., Rodda, J.C. (2003). World water resources at the beginning of the 21st century. Cambridge: UNESCO and Cambridge University Press.

Sinclair, J. R. (2019). Importance of a One Health approach in advancing global health security and the Sustainable Development Goals. *Revue scientifique et technique (International Office of Epizootics)*, **38**(1): 145-154.

Strosnider, H., Azziz-Baumgartner, E., Banziger, M., Bhat, R.V., Breiman, R., Brune, M.N., DeCock K., Dilley, A., Groopman, J, Hell, K., Henry, S.H. (2006). Workgroup report: public health strategies for reducing aflatoxin exposure in developing countries. *Environmental Health Perspectives*, **114**(12):1898–1903.

The Times of India (2021). National Institute of One Health to come up on Mafsu campus. Web: https://timesofindia.indiatimes.com/city/nagpur/natl-institute-of-one-healthto-come-up-on-mafsu-campus/articleshow/70899851.cm [Accessed on 12th March, 2021].

Thompson, R. (2013). Parasite zoonoses and wildlife: One health, spill over and human activity. *International Journal for Parasitology*, **43**(12-13): 1079-1088.

Thompson, R.C.A., Lymbery, A.J., and Smith, A. (2010a). Parasites, emerging disease and wildlife conservation. *International Journal for Parasitology*, **40**(10): 1163–1170.

Thrusfield, M. (1995). *Veterinary Epidemiology*. 2nd Edn. Blackwell Science Ltd, Oxford.

U.S. DEPARTMENT OF AGRICULTURE (2015). One Health. Web: https://www.usda.gov/topics/animals/one-health [Accessed on 31st January, 2023]

Vector Biology and Control [meeting held in Geneva from 15 to 21 September 1987]. World Health Organization.

Veterinary Council of India (2019). Minimum Standards of Veterinary Education. Available from: https://www.kcvas.com/images/new%20 syllabus.pdf. Retrieved on 12-06-2020.

Wassie, S. B. (2020). Natural resource degradation tendencies in Ethiopia: A review. *Environmental systems research*, **9**(33): 1-29.

Watson, M.J. (2013). The costs of parasites-What drives population-level effects?

Meta-analysis meets life-history. *International Journal for Parasitology: Parasites and Wildlife* **2**(1): 190–196.

Wayne, A.F., Maxwell, M., Ward, C.G., Vellios, C.V., Ward, B., Liddelow, G.L., Wilson, I., Wayne, J.C., and Williams, M.R. (2013b). The importance of getting the numbers right: quantifying the rapid and substantial decline of an abundant marsupial, *Bettongia penicillata*, *Wildlife Research*, **40**(3):169-183.

White, A. and Hughes, J.M. (2019). Critical importance of a one health approach to antimicrobial resistance. *Ecohealth*, **16**(3): 404-409.

World Health Organization (1988). WHO Expert Committee on Vector Biology and Control, and World Health Organization. *Urban vector and pest control: 11ᵗʰ report of the WHO Expert Committee on* 'Vector Biology and Control' [meeting held in Geneva from 15 to 21 September 1987.

WHO India. (2018). World Health Organization. Union Government Launches a New State-of-the Art Information Platform to Monitor Public Health Surveillance. Available from: http://www.origin.searo.who.int/india/areas/health_systems/ihip_launch_2018/en/#:~:text=ms%20preeti%20sudan%2c%20union%20health,ihip)%20in%20seven%20states%20today.&text=the%20first%20one%2dof%2dits,technologies%20and%20digital%20health%20initiatives. Retrieved on 16-06-2020.

World Health Organization (2002). WHO Study Group on Future Trends in Veterinary Public Health Meeting, and World Health Organization. Future Trends in Veterinary Public Health: A Report of a WHO Study Group.

World Health Organization (1975). The veterinarian contribution to public health practice. Report of a joint FAO/WHO expert committee on veterinary public health. WHO, Geneva, Technical Report Series No. 573, pp. 79.

World Health Organization (2012). Technical Consultation: Strategies for Global Surveillance of Antimicrobial Resistance. Web: https://www.who.int/iris/bitstream/10665/90975/1 WHO_HSE_PED_2013.10358_eng.pdf [Accessed on 3rd March, 2023].

World Health Organization (2014). Antimicrobial Resistance: Global Report on Surveillance. Web: https://apps.who.int/iris/bitstream/handle/10665/112642/ 9789241564748_eng.pdf [Accessed on 3rd March, 2023].

World Health Organization (2017). One Health Web: https://www.who.int/news-room/questions-and-answers/item/one-health [Accessed on 9th March, 2023].

World Health Organization (2022). One health Web: https://www.who.int/news-room/fact-sheets/detail/one-health [Accessed on 9th March, 2023].

World Health Organization Country Office for India (2021). National action plan on antimicrobial resistance. Web: https://ncdc.gov.in/WriteReadData/l892s/File645.pdf [Accessed on 3rd March, 2021].

World Organization for Animal Health (2023). One Health. Web: https://www.woah.org/en/what-we-do/global-initiatives/one-health/#ui-id-4 [Accessed on 9th March, 2023].

Wyatt, K.B., Campos, P.F., Gilbert, M.T.P., Kolokotronis, S.-O., Hynes, W.H., DeSalle, R., Daszak, P., MacPhee, R.D.E., and Greenwood, A.D. (2008). Historical mammal extinction on Christmas Island (Indian Ocean) correlates with introduced infectious disease. *PLoS ONE*, **4**(1): 1–9.

Yadouleton A.W.M., Asidi A., Djouaka R.F., Braïma J, Agossou C.D., Akogbeto M.C. (2009). Development of vegetable farming: a cause of the emergence of insecticide resistance in populations of *Anopheles gambiae* in urban areas of Benin. *Malaria Journal*, **8**(103):1–8.

Yan, Z., Xiong, C., Liu, H., & Singh, B. (2022). Sustainable agricultural practices contribute significantly to One Health. *Journal of Sustainable Agriculture and Environment*, **1**(12): 165-176.

Printed in the United States
by Baker & Taylor Publisher Services